The Oil Game

Also by James McGovern

FICTION

Fräulein

No Ruined Castles

The Berlin Couriers

NONFICTION

Crossbow and Overcast

Martin Bormann

To the Yalu

To Give the Love of Christ:
A Portrait of Mother Teresa and the
Missionaries of Charity

The Oil Game

James McGovern

The Viking Press
New York

First published in 1981 by The Viking Press
625 Madison Avenue, New York, N.Y. 10022
Published simultaneously in Canada by
Penguin Books Canada Limited

LIBRARY OF CONGRESS CATALOGING IN PUBLICATION DATA
McGovern, James.
The oil game.
1. Petroleum industry and trade—United States.
I. Title.
HD9565.M26 338.2'7282'0973 81-65270
ISBN 0-670-52134-5 AACR2

Grateful acknowledgment is made to the following for permission to reprint copyrighted material:

Dow Jones & Company, Inc.: Selections from the January 30, 1980, issue of *The Wall Street Journal*, "The Bookshelf" ("Takeover") section, by David Sanford. Copyright © 1980 by Dow Jones & Company, Inc.
Foreign Affairs: A selection from "Oil and the Decline of the West" by Walter J. Levy, from the Summer 1980 issue of *Foreign Affairs*.
Los Angeles Times: Selections from the article "Dispute Over Rules Not a Soccer Game—Bloom" by Martin Baron, from the February 10, 1980, issue of the *Los Angeles Times*. Copyright © 1980 by the Los Angeles Times.
The New York Times Company: Selections from "Economic Scene" by Leonard Silk, which appeared in the June 11, 1980, issue of *The New York Times*; selections from an article by Senator Howard Metzenbaum, which appeared in the September 16, 1979, issue of *The New York Times*; selections from an article by Walter J. Levy, which appeared in the January 4, 1979, issue of *The New York Times*. Copyright © 1979, 1980 by The New York Times Company.

All "Letters to the Editor" are reprinted by permission.

Printed in the United States of America
Set in Linotron Primer
Second printing August 1981

To Elizabeth McGovern, daughter, and
Adam Kennedy, friend

Foreword

In 1979 Exxon became the first industrial corporation in the world to earn more than $4 billion in a single year. For all other oil companies, it was also a very good year. But for people who had to buy their products, 1979 was not quite so good.

You may have been among those who fumed on long gas station lines that summer, wondering why gasoline was in short supply as the price climbed to well over $1 a gallon. You may have been outraged as the price of No. 2 home heating oil approached $1 a gallon. You may have complained as inflation, which is directly traceable to rising energy prices, continued to take bigger bites out of your pay or pension check.

The oil companies had no cause for complaint. The profits of the eight biggest American companies came to $15 billion. Four years earlier their combined profits had been $6.8 billion. House Speaker Thomas ("Tip") O'Neill termed such profits "sinful," while James Archuleta, head of the Oil, Chemical and Atomic Workers, singled out Exxon's as "pornographic." Nonetheless industry leaders made no apologies to irate legislators and a skeptical public about their awesome profits. Instead, they contended that most of their gains resulted from a surge in foreign earnings, and insisted that they needed the money to find more oil and gas to avert future shortages; consumers would ultimately benefit from the higher prices, they said.

Maurice F. Granville, chairman of Texaco, summed up the oil

company position in a report to his stockholders: "Texaco expects to play an important role in satisfying future energy demands, but to do so it must have earnings high enough to support the enormous requirements for capital investment." Mr. Granville countered the suspicion that Texaco was profiteering from the energy crisis with the observation: "Despite Texaco's increased level of earnings in 1979, the Company's average profit per gallon on all petroleum and products sold worldwide during 1979 was only 2.5 cents per gallon."

Where does the truth lie? What, you are entitled to wonder, is going on here? Surely something strange, possibly even sinister, when total profits of the leading twenty-five oil companies have increased nearly 300 percent since 1972, the last full year of cheap energy, while gasoline and heating oil prices continue to soar. What are the oil companies doing with the enormous amounts of cash pouring into their coffers? Is there an energy crisis that will mean future shortages, or is the whole thing simply a conspiracy to boost industry profits? After all, there were gasoline shortages in the summer of 1979, but there was plenty of gasoline available at higher prices just a year later.

Yes, there is an energy crisis. But no, you may be relieved to learn, this is *not* still another book about the many ramifications of the energy crisis or the prospects for coal, nuclear, solar, and hydro power, synthetic fuels, windmills, and other alternate energy sources. Its subject is the American oil business, how it operates today, and how those operations affect you.

Contents

Introduction

The security, prosperity, and sanity of the world's industrial democracies are all dependent on one substance: petroleum. The word "petroleum" comes from the Latin *petra* (rock) and *oleum* (oil)—the literal meaning is oil from rocks. Petroleum is an organic mineral formed by the alteration of the remains of prehistoric animals and plants and is by no means new to mankind. The Chinese drilled for oil as early as the third century B.C. using bamboo tubes and bronze bits. Belisarius, a Greek general of the later Roman Empire, is reported to have broken the ranks of the Vandals in North Africa by greasing a herd of pigs with oil, setting them afire, and sending them rushing at the enemy. When Alexander the Great visited Ecbatana near Baku (now an important Russian oil center), he was entertained by a display of burning oil and the Eternal Fires.

The mixture of thousands of different hydrocarbons (compounds of hydrogen and carbon) that is petroleum becomes commercially valuable and economically useful only after it has been processed in a refinery into products like gasoline, heating and fuel oil, diesel fuel, and jet fuels. But without it, the economies of the industrialized democracies would come to a dead halt. So it is worth taking a hard look at the oil game, a game where the financial stakes are immeasurably high for the American economy and the national security (the Navy, Air Force, and the combat forces of the Army—except for its footsloggers—mostly run on oil).

The domestic and international operations of Big Oil are little known and make for a revealing and intriguing tale. The petroleum industry occupies center stage in the energy drama, for it supplies three-quarters of the energy America consumes today—either by producing it domestically or by importing it from abroad. It will continue to play the key role at least through the end of this century, since it just might make some big "plays" that will arrest the decline in U.S. oil production and is maneuvering to become dominant in alternate fuels, especially coal and synfuels.

To understand the oil game, it is necessary to realize that oil companies, although they look pretty much alike to the outsider, upon close inspection exhibit differences in almost everything except a shared devotion to the bottom line. They range all the way from Exxon, whose gross revenues of $84.35 billion in 1979 were more than the gross national product of Sweden, to an independent wildcatter with a two-room office over a grocery store in Ponca City, Oklahoma.

The Seven Sisters, whose rise was so ably chronicled by Anthony Sampson in his book of that name published in 1975, still rule the international oil game, although their rule is not quite so autocratic as it was when his book appeared. Ranked in order of 1979 revenues, the Seven Sisters are Exxon, Shell, Mobil, Texaco, British Petroleum, Standard Oil of California, and Gulf. Shell is the Royal Dutch/Shell Group, based in The Hague, the Netherlands, and owned by two private companies, the Royal Dutch Petroleum Company of The Hague (60.7 percent) and the "Shell" Transport and Trading Company Ltd. of London (39.3 percent). Royal Dutch/Shell operates in the United States through the Houston-based Shell Oil Company, itself the thirteenth largest U.S. industrial corporation as ranked by sales. The British Petroleum Company (originally the Anglo-Persian Oil Company) is based in London and is 51 percent owned by the British government, although it has operational freedom from the politicans in Whitehall. BP holds a 53 percent interest in Standard Oil of Ohio. The other five Sisters are totally American owned.

The Seven Sisters are "major" oil companies. They are also

known as "integrated multinationals"—*integrated* because they engage in all the industry's successive stages of operations: exploration, crude oil production, refining, transportation, and marketing; and *multinational* because they conduct these operations in almost every country in the noncommunist world.

Ranking just below the five American-owned Sisters among the top eight U.S. "majors" in sales are Standard Oil of Indiana (Amoco) of Chicago; Atlantic Richfield (Arco) of Los Angeles; and Conoco (formerly Continental Oil) of Stamford, Connecticut. None of these three, however, plays a significant role on the international scene.

Checking in below the eight U.S. majors are the "lesser majors" or "independents." These are also integrated companies, but occupy a relatively minor position in the world industry. Ranked by 1979 sales, they are: Sun (Radnor, Pennsylvania); Occidental Petroleum (Los Angeles); Phillips Petroleum (Bartlesville, Oklahoma); Standard Oil of Ohio (Cleveland); Union Oil of California (Los Angeles); Amerada Hess (New York); Marathon Oil (Findlay, Ohio); Ashland Oil (Russell, Kentucky); Cities Service (Tulsa, Oklahoma); Getty Oil (Los Angeles); and Charter (Jacksonville, Florida).

Below this elite group in size and influence are the nonintegrated independents, companies that are limited to only one of the successive stages of operations. For example, a company might be an independent refiner that buys crude oil from others and in turn sells the finished product to wholesale and retail cooperatives, agricultural accounts, industrial and commercial accounts, government agencies, and retail dealers in non-brand-name gasoline. Still further down the scale in the nonintegrated segment of the business are the brokers and jobbers who arrange these deals.

Then there is that hardy, risk-taking entrepreneur known as the independent wildcatter who drills for new oil with the backing of banks, private investors, and large corporations. Wildcatters average about one hit in ten attempts, though a few of these ten thousand independents do considerably better. For instance, the small New Orleans–based McMoRan Oil and Gas

Company had an unusual 60 percent success rate in 1978 and 1979. For the fiscal year ending June 30, 1979, it had revenues of $19.4 million and a net income of $1.01 million—not bad for a small independent wildcatter, although such figures pale by comparison with the operations of the integrated multinational Seven Sisters.

Standard Oil of California (Socal), based in San Francisco, racked up worldwide sales of almost $30 billion in 1979, making it the sixth largest U.S. industrial corporation by that measurement (after Exxon, General Motors, Mobil, Ford, and Texaco). Socal found oil in California, the Gulf of Mexico, Wyoming, and Alaska, and imported crude for its own refineries from Nigeria, Bahrain, Venezuela, Saudi Arabia (where it was the first to discover oil), and Indonesia, where it shares a fifty-fifty partnership with Texaco in Caltex. Its 1980 assets were $7.1 billion, with more than $6 billion represented by its principal subsidiary, Chevron U.S.A., which had nineteen thousand employees and provided 45 percent of the corporation's profits.

Now this is big business of a magnitude almost impossible to comprehend, but Texaco (not to mention Exxon) is bigger still. Texaco's 1979 worldwide sales were almost $39 billion, making it the fifth largest U.S. industrial corporation by that measure. It marketed more than eight hundred petroleum products and five hundred petrochemicals through its companies in the United States, Canada, Latin America, Europe, and West Africa and through nonsubsidiary companies elsewhere in Africa, the Middle East, the Far East, and Australasia. Its products range from Sky Chief gasoline to paving asphalts to rust inhibitors. Like the other Sisters, Texaco explores for and produces petroleum, owns pipelines, refineries, and tankers, and markets everywhere from your friendly neighborhood station to Ecuador (Lubricantes y Tambores del Ecuador, C.A.), Spain, the Canary Islands (Texaco Canarias S.A.), and Mali (Société Malienne des Pétroles Texaco).

These heady figures and farflung global activities are hard to digest, but the name of the game in the oil business is rather simple: access to crude oil. It is their unparalleled worldwide access to crude that gives the Seven Sisters their dominant position, but this access comes at a high price to Americans and the supplies are located to a disturbing degree in the increasingly unpredictable Middle East, which makes the seemingly gilt-edged bonds of the international companies less secure than their financial statements suggest. Exxon obtained 57 percent of its supply from the Middle East in 1978, only 19 percent from the United States, and the remainder (35 percent) from other regions, principally Latin America, Africa, and the Far East. Here are the sources of crude oil for the other four American sisters in 1978:

Company	*Mideast*	*United States*	*Other*
Gulf	45%	25%	30%
Mobil	63%	17%	20%
Socal	71%	12%	17%
Texaco	67%	15%	18%

How did the United States, which built the greatest industrial machine the world has ever seen, arrive at the point where its entire lifestyle is tied to the goodwill of a few Mideast autocrats whose lands were blessed by Allah—or, more prosaically, by an accident of geology that gave them some of the world's largest oil deposits?

The Oil Game

Chapter One

The Business

The president, a canny political veteran whom some considered colorless and overly cautious, could have told the American people the simple truth—that they were running out of oil. But he said nothing.

The president was James Buchanan. The year was 1859, when Colonel Edwin Drake drilled the first well specifically for the purpose of finding oil in Titusville, Pennsylvania. Drake's success effectively began the world's biggest, and most misunderstood, industry, and one that remained until recently dominated by American companies and American citizens.

Although he was called "Colonel," Drake held no military rank. He was a frail, bearded thirty-eight-year-old retired conductor on the New Haven Railroad who was tortured by neuralgia. He was sent to Titusville, an area where oil had been seeping out of the ground at least since the time of the Seneca Indians, by a group of businessmen who had formed the Pennsylvania Rock Oil Company. Their purpose was to find large quantities of crude oil that could be refined into kerosene and sold as lamp fuel in place of dwindling supplies of whale oil.

Why the promoters chose Drake to lead their field operations is a mystery, for he had neither business experience nor technical qualifications. But then, nobody else knew much about drilling, Drake was unemployed and could be hired cheaply, and he had a railroad pass that could get him to Pennsylvania for nothing. The promoters' initial investment was $1,000.

Once on the banks of Oil Creek in western Pennsylvania, the quietly persistent Drake became the object of some ridicule. His drilling rig—an iron bit attached by a rope to a wooden windlass—became known as "Drake's Folly." But after months of failure, the bit was withdrawn from the well at 69½ feet and a dark green liquid seeped to a few feet below the surface.

Thus the petroleum age was born, and simultaneously began to die. For there is one central fact that must be understood about the oil business: Oil is a finite resource. Once discovered and produced, it immediately begins to disappear. It cannot be replaced, like apples on a tree or corn in a field. The absolute certainty is that it will vanish entirely in the United States, and in the rest of the world, in one of the coming decades. The only question is when.

An oil boom resembling the California gold rush followed Drake's discovery. Speculators, promoters, gamblers, lease grabbers, people who said they could smell out oil in the ground, teamsters, and adventurers, including John Wilkes Booth, flocked to western Pennsylvania. The hills echoed to the pounding of thousands of drill bits. Shantytowns mushroomed. Refineries and pipelines were built, railroad tank cars developed. The price of oil fluctuated wildly, from $20 a barrel in 1860 to 10 cents a barrel in 1861. Coopers toiled furiously making barrels for the oil. Those wooden barrels, holding 42 gallons, remain the world's chief measure of oil.

Kerosene soon replaced tallow candles and whale oil as a source of light for Americans, and was sold as far abroad as China. Oil lubricated machinery and fueled stoves and ships. But what of the man who started it all? It is a truism of the oil game that discoverers and innovators rarely get rich; that fate falls to their financial backers. The entrepreneurs of the Pennsylvania Rock Oil Company grew wealthy, but they shunted "Colonel" Drake aside. He died in abject poverty, an embittered man. However, he was rescued from obscurity when a Standard Oil partner had his body placed under a $100,000 monument in Titusville. A Drake Well Museum exists in that now quiet town.

The production of crude oil moved westward and increased

steadily in the United States for more than a century. In 1892, Edward Laurence Doheny drilled Los Angeles's first well. Like Colonel Drake, he was a neophyte who operated on a shoestring.

Doheny used a pick and shovel to dig a 4-by-6-foot mining shaft and the hollowed-out trunk of a eucalyptus tree as a drill. He came up with 45 barrels a day at Beverly and Glendale boulevards, and five years later, there were twenty-five hundred wells and two hundred oil companies in Los Angeles. A forest of wooden derricks rose over the city. The oil, used mainly for axle grease and refined into kerosene, was so free-flowing and cheap that the Los Angeles City Council had it sprayed on unpaved roads to hold down the dust. By the early 1920s, southern California was producing 25 percent of the world's oil output.

In 1901, the Spindletop field was brought in near Beaumont, causing Texas to begin to challenge California and Oklahoma as the largest producing state. (Texas finally took first place in 1928.) It has been said that Spindletop and the selection of a native, Debra Jo Fondren, as *Playboy* cover girl and Playmate of the Year for 1978 are the only two things that ever happened to Beaumont.

Spindletop, a 10-foot mound from which oil often seeped, was the dream of a one-armed fanatic named Patillo Higgins, son of a Beaumont gunsmith. Many of his fellow townspeople considered him a madman because of his decade-long insistence that there was oil under that mound near the Gulf Coast.

One of those who did not was Captain Anthony Lucas, a native of Austria who had graduated as an engineer from the Polytechnic Institute at Graz and drilled for salt in Louisiana. Intrigued by Higgins's claims, Lucas brought him his technical expertise and some modest financial aid. When Lucas exhausted his savings and credit after months of drilling failures, he was obliged to seek backing from Standard Oil.

Calvin Payne, a Standard Oil expert of worldwide experience, reported to his headquarters: "No indication whatever to warrant the expectation of an oil field on the prairies of southeastern Texas." It was one of the few mistakes Standard Oil ever made, and certainly the most costly. Lucas later obtained the backing of other eastern capital, principally from the Mellons of Pittsburgh, and brought in the biggest gusher ever seen on the American continent—the cannonade of its first eruption caused cattle 4 miles away to stampede. Spindletop was the genesis of what were to become Sun Oil, Gulf Oil, and Texaco. It broke the worldwide monopoly of Standard Oil, although Standard later did buy into the field by obtaining a controlling interest in the Texas-based Humble Oil. Patillo Higgins and Captain Lucas profited only modestly from their discovery, having been bought out early by partners with more capital.

Spindletop became an extravaganza of wealth and waste. Conservation was then hardly thought of, much less practiced. Each producer took out as much oil as he could, as fast as he could. Overdrilled, overproduced, ravaged by fires because of lax safety measures, the great field petered out by 1905.

No matter. The prevailing conviction was that America would never run out of oil, that there would always be other Spindletops. Indeed other strikes followed along the Gulf Coast, in Colorado, Oklahoma, Louisiana, California, and Kansas. By 1909, the fiftieth anniversary of Titusville, the U.S. production of oil was half a million barrels daily, greater than that of the rest of the world combined.

Most U.S. production fell under the centralized control of the Standard Oil Trust, formed by John D. Rockefeller in 1882, which also operated overseas. Standard Oil, through its numerous subsidiaries, controlled 90 percent of domestic refining capacity and the transportation systems—pipelines and railroads—that fed the refineries. With such a commanding position, it was able to regulate the production of petroleum and set the price of petroleum products on the American market. But its cold-blooded monopolistic practices and burgeoning wealth drew the hostility of populist politicians, the Hearst press, muckraking journalists like Ida Tarbell, ordinary citizens, and the governors and attorney generals of such states as Ohio and Texas. Standard Oil's blatant influencing of legislators combined with the secrecy with which it conducted its interlocking operations (a secrecy characteristic of all big oil companies even today, rationalized on the grounds of protecting "proprietary information") created outright rage. Big Oil was accused of being a conspiracy bent on manipulating the market for the sake of its own profits, an accusation that has persisted to this day.

Finally, in response to the complaints, the Supreme Court dissolved the Standard Oil trust in 1911, breaking it up into thirty-four separate companies and opening the way for the growth of other majors like the Texas Company (now Texaco) and Gulf. Nevertheless, the direct descendants of the splintered trust have not fared badly. Standard Oil of New Jersey, now Exxon Corporation, is not just the world's biggest oil company but the world's biggest company. Standard Oil of New York (now Mobil), Standard Oil of California (Socal, marketing under the Chevron brand), Standard Oil of Indiana (marketing as Amoco), and Standard Oil of Ohio (selling gasoline as Sohio in Ohio, Boron in neighboring states, and BP in the East) all rank among the thirteen top U.S. oil companies.

In 1911, however, the financial outlook for the offshoot companies was not so bright. In fact, the future of the entire oil business was in doubt as Edison's electric lamp began supplanting the gas lamp. Then came the practical development of the internal combustion engine. In 1911, there were fewer than half a million "horseless carriages" in the United States. By 1920, there were nine million and the oil industry had a vast new market for what had once been a useless by-product of the distillation process—gasoline.

The assets of the majors just about doubled in the 1920s. To the citizens of the world's largest producer, unlimited supplies of cheap oil seemed a birthright, a mandate from heaven that would never pass. But of course it had to, although few in the industry or government said so publicly. One body that did was the Federal Oil Conservation Board, which warned in 1929:

> According to the present opinion of our best petroleum geologists, our total resources, instead of being 68 percent of those of the world, are not more than 18 percent. If our petroleum reserves are not to be drawn upon at a faster rate than those of

> other countries, our resources should be several times larger.
>
> The obvious inference is that the United States is exhausting its petroleum reserves at a dangerous rate. If the international comparison is made, this country is depleting its supply several times faster than the rest of the world. How real is the danger expressed in this fact, and what remedy can be devised are questions confronting the American people as they plan for the future. . . . The depletion rate of our own resources can be brought more into accord with that of foreign resources in only one way—importing a greater quantity of crude petroleum.

Few paid any heed to this warning. The American people and their elected representatives continued to behave as though there were no tomorrow where oil was concerned. One American who definitely did not believe that his country was running out of oil was Columbus Marion Joiner.

"Dad" Joiner was in 1930 a small, spry, bright-eyed man of seventy with a winning manner who radiated honesty. Although he had only seven weeks of formal schooling, he liked to read books and could quote long passages from the Bible and Shakespeare. Born on a farm in Alabama, he engaged in a variety of occupations before becoming a wildcatter in Oklahoma in 1913, at the age of fifty-three. A wildcatter obtains leases on ground where he thinks there might be oil and persuades others to back him in drilling for it. When successful (most of them are not), the wildcatter is transformed into an oil man. Strangely, it is wildcatters, not the major oil companies with their enormous resources, who have found most of the big oil fields in the United States; the breed is still doing most of the exploring.

After making a few dollars but enduring a lot of bad luck in Oklahoma, Dad Joiner convinced himself that oil could be found in the dirt-poor farmland of east Texas in the basin west of the Sabine River. His optimism was bolstered by the recommendations of his "geologist," A. D. "Doc" Lloyd, a gregarious six-foot-tall, 320-pound Oklahoman. Doc Lloyd was neither a doctor nor a geologist and his real name was Joseph Idelbert Durham. In the course of a varied career, he had worked as a drugstore clerk, self-taught mining engineer and government chemist examining ore samples during a gold rush in Idaho, seller of patent medicines, and adviser to Dad Joiner in previous futile wildcatting ventures.

Doc Lloyd surveyed the few leases Dad Joiner had secured and concluded that there was oil beneath the sweet potato, corn, and cotton fields of east Texas. Dad Joiner circulated his report and map to banks and major oil companies in an effort to raise the financial backing he needed to proceed.

The geologists of the major oil companies flatly disagreed with Doc Lloyd; they could find none of the usual geographical indications of potential oil wells in east Texas. One expert, W. Dow Hamm, later to become chief geologist of Atlantic Richfield, commented: "I didn't have the heart to tell Joiner how worthless this map was."

Nevertheless, Dad Joiner persisted. He scraped together some money from local farmers and bought up leases in a complicated series of deals. In August 1927, with a rickety, worn-out rig, he began what was known as a "poor boy" operation on the 975½-acre farm of the genial, cooperative widow Daisy Bradford.

At first it seemed the big oil companies were right. Daisy Bradford No. 1 was a dry hole, and so was No. 2. Although he was running short of money and his equipment was falling apart, Dad Joiner ordered an attempt on Daisy Bradford No. 3 in May 1929 with a new head driller, Ed C. Laster. In the summer of 1930, Laster came up with some sand that contained oil. He told Dad Joiner they should set up piping, casing, and other equipment to bring in Daisy Bradford No. 3 as a producing well. Laster then informed a local newspaper, the Henderson *Times:* "I believe we have the biggest thing yet found in Texas."

The attempt to bring in the well was begun in early October. More than five thousand people, many of whom had invested $25 in Joiner's venture, flocked to the Bradford farm to watch. Among them were scouts from the major oil companies and Haroldson Lafayette Hunt, an independent wildcatter who owned about a hundred wells in Arkansas and Louisiana and had already amassed a substantial bankroll.

Late on the afternoon of October 5, a swab—a device of rubber and steel that creates suction—was lowered 3,500 feet into the hole and brought up oily sand. Toward nightfall, Joiner and Laster heard a gurgling noise rising from the well. "Put out your cigarettes! Quick. Douse the boilers!" the head driller shouted.

Mud, sand, and gas roared out of the well and then a stream of oil shot over the derrick, spraying the spectators, who began to jump up and down, shout, dance, throw their hats in the air, and fire pistols. Leaning against the derrick, a dead-tired Dad Joiner told Laster: "I always dreamed it but I never believed it."

What Dad Joiner had found became—after others rushed in to exploit it fully—the biggest oil field in the world at the time. One vast pool of oil, covering 140,000 acres, stretched 5 to 12 miles from east to west and 45 miles from north to south.

Unfortunately, his discovery brought only trouble to Dad Joiner. He was besieged by creditors and investors, not all of whom he was able to satisfy. Many of his leases turned out to stand on murky legal ground, he became embroiled in over 150 lawsuits, and a receivership was appointed for his operation. It was then that H. L. Hunt, who had taken a chance on other leases in east Texas, stepped in with an offer to help. With borrowed money, Hunt proposed to buy Joiner's leases, covering about 4,000 acres, lawsuits and all. "Boy, you would be buying a pig in a poke!" Joiner told him, but agreed to the deal.

The money Dad Joiner got out of his discovery went for legal fees and futile attempts to find another Daisy Bradford No. 3. In March 1947, at the age of eighty-seven, he died in his modest home at 4637 Mockingbird Lane in Dallas. Unlike Colonel Drake, he never displayed any bitterness at the hand fate had dealt him, although, like Drake and Doc Lloyd, who passed away in a Chicago hotel room in 1941, Dad Joiner died broke.

H. L. Hunt made about $100 million on the gamble he took with Dad Joiner's leases. That profit served as the stake that eventually made him possibly the richest man in America, with a fortune estimated at more than $2 billion. Unlike the independent Hunt, the major oil companies were slow to act in east Texas. However, by 1938, they had gained control of an estimated 80 percent of the wells and proven acreage there by buying or leasing land.

This story illustrates why the oil business is so different in the United States from what it is in other countries. In this country, a property owner not only owns the surface land but also the subsoil mineral rights below it. In order to extract the crude, the oil companies, which are also privately owned, must negotiate with individual property owners (who can number in the thousands) to buy or lease the land; the standard landowner's royalty is one-eighth. In other countries, governments own the subsoil rights. They either explore themselves with nationalized companies or lease exploration rights to a few favored private companies. For example, the British government owns the oil in the North Sea, and the Saudi Arabian government (in effect, the Saudi royal family) owns all the vast reserves in Saudi Arabia—the desert bedouins mounted on camels own nothing.

The bonanza of the east Texas field made millionaires of H. L. Hunt and a lot of poor farmers, but it proved to be a mixed blessing for the oil companies in the Depression years of the early 1930s. To understand the oil business, a central fact must be grasped. The oil companies don't want shortages and long lines at the gas pump. They don't deliberately create shortages, for they want enough product to sell at a profit and as little government intervention in their affairs as possible. But neither do the companies want a situation where there is too much oil, for when supply overwhelms demand, prices fall. This is a blessing for consumers, of course, but bad for company profits. This truism is as valid today as it was in the early 1930s, when it was illustrated in classic fashion by the east Texas experience.

Combined with new discoveries in Oklahoma and South America, the giant east Texas field created a glut on the market. The price of crude oil fell from $1.30 a barrel in 1930 to 35 cents a barrel, then 20 cents. In July 1931, it had plummeted to 10 cents, with some large shipments selling for 5 cents a barrel. Oil was, in fact, cheaper than fresh water, which sold for 10 to 25 cents a barrel because of the cost of hauling.

Those who claim the oil companies need huge profits to underwrite an all-out effort to increase domestic production should be reminded that independent wildcatters have done most of the discovering in the past and that overproduction has always been the chief threat to the companies' financial position. It was only when the federal government and the state of Texas stepped in to regulate production in east Texas that the balance between supply and demand (and company profits) was restored.

Even with regulation there was so much oil that a "hot oil" war ensued. Risking the wrath of the Texas Rangers, independent operators produced and sold oil illegally, just as bootleggers had sold booze a few years earlier. Given this situation, who could believe that the United States was running out of oil? But implacably, year by year, it was and the signs were there to be read.

In 1948, although the United States produced half the world's oil, it imported more crude oil than it exported. It was the first year that this had happened. Most of the imported oil came from Venezuela, with some starting to flow in from the Middle East; in both areas, oil was much cheaper to produce than it was in the United States.

In March 1956, in an event hardly noticed by the media and the public, Dr. M. King Hubbert addressed a meeting of petroleum engineers at the Southwest Section of the American Petroleum Institute in San Antonio. Born in 1903 in central Texas, Hubbert had received B.S., M.S., and Ph.D. degrees from the University of Chicago, taught at Columbia for ten years, and was a research geophysicist for Shell while holding a visiting professorship at Stanford. He was something of a maverick loath to accept the conventional wisdom, and had for some time been studying and charting the cycles of the discovery, exploitation, and depletion of nonrenewable resources, particularly petroleum.

His address shook up the assembled members of the American Petroleum Institute who, as leaders of an industry that was the cornerstone of the economy and had been prospering and expanding since the time of Colonel Drake, had every apparent reason to be confident of a rosy future. With graphs, charts, and impeccable logic, Dr. Hubbert demonstrated that U.S. crude oil production would peak between 1966 and 1971 and decline thereafter, never to resume an upward curve. If Dr. Hubbert's analysis was correct, his audience knew it was the beginning of the end of the petroleum industry and the American lifestyle so dependent on it.

The industry hoped that he was wrong and underwrote studies to prove that he was. Robert Dunlop, president of Sun Oil Company, commented that "reports of the U.S. Geological Survey show that we are a long way from exhausting our petroleum resources . . . enormous supplies are still available for development." As Dr. Hubbert observed: "The universal intuitive judgment of the U.S. petroleum industry at that time was that there was plenty of oil left. The propaganda dictum of the A.P.I. was that the United States has all the oil it needs."

This propaganda was in character, for the U.S. companies were extracting substantial tax benefits and other incentives from the government based on the proposition that the United States still possessed enormous reserves and the companies needed the inducement of preferential tax treatment to explore for new oil. The continued insistence that U.S. production would not decline also enabled the oil companies to beat back demands to abolish the oil import quota—a position that could only be justified if domestic reserves were indeed adequate to satisfy future American requirements.

This difference of opinion between Dr. Hubbert—who could hardly be considered an eccentric or a foe of the oil companies—and the industry created no public stir. But it had great import for the major oil companies and their profit margins.

After 1956, low-cost foreign crude became increasingly available. Since most domestic production was controlled by the major oil companies, the relatively smaller companies and the larger independents went abroad to seek oil, which they imported into the United States, the largest market. It cost only a few cents a barrel to produce oil in the Middle East. Even with tanker costs and a modest impost duty of 10 cents a barrel, the cheaply produced foreign crude was underselling domestic crude on the eastern seaboard as early as the mid-fifties.

The inevitable result was reduced profits for the big U.S. producers. By 1958, the rates of return had dropped 45 percent for Shell and Mobil and 43 percent for Exxon. The federal government obligingly intervened to stabilize the market by instituting a voluntary import quota program on July 1, 1957. When this did not prove entirely satisfactory because some independents refused to cooperate, a mandatory quota was established on March 11, 1959.

"National security" was the rationale offered for the mandatory program, which remained in effect for fourteen years. But, of course, it also protected a politically powerful special interest group. By restricting imports of foreign crude, which in those pre-OPEC days was both cheap and plentiful, higher prices and oil company profits were maintained.

This stability had less pleasant effects for everyone but the oil companies. It cost the American consumer an estimated $4 billion a year in the early 1960s, according to Professor M. A. Adelman, an economist at MIT. It placed American industry at a disadvantage as the low-cost crude flowed to foreign manufacturing competitors. The quotas also antagonized the producing nations, planting the seed for the formation of OPEC. Most serious of all, by forcing the world's largest market to rely on domestic production to supply almost 90 percent of its needs, the program severely depleted U.S. reserves.

The price had to be paid, and it was. In 1970, U.S. production reached 11.3 million barrels a day. That was its peak; even with the Alaskan discoveries, it has been declining ever since. King Hubbert had been right.

As domestic production declined, demand continued to increase. This imbalance was noted by some canny officials in Venezuela and the Middle East. They began to think that if they formed a cartel and fixed the production and price of their oil, in imitation of the old Standard Oil trust, the asset they had been almost giving away would be worth a lot more on the U.S. and world market.

Juan Pablo Pérez Alfonso, oil minister of Venezuela, led the effort to form an international association of oil producers, a cause approved by the oil minister of Saudi Arabia. The giant international oil companies had long been dictating the price of the producer countries' oil. Because of a glut on the world market in August 1960, the companies, led by Exxon, sharply dropped their posted price. This provided the catalyst that rallied other producers to the cause of Venezuela and Saudi Arabia. At a meeting in Baghdad, Iraq, in September 1960, they were joined by Iran, Iraq, and Kuwait in forming the Organization of Petroleum Exporting Countries. Eight other nations later joined OPEC.

A government cartel had been formed to counter a private cartel. But in its early years, OPEC, because of its members' conflicting interests, failed to dent the unified front of the multinational Seven Sisters (Exxon, Mobil, Texaco, Gulf, Socal, Royal Dutch/Shell, and British Petroleum), who successfully played off one government against the other.

Chapter Two

The Seven Sisters Meet Big Brother

In 1973, as U.S. demand continued to increase and production to decrease, the Nixon administration abandoned import quotas entirely. Foreign oil was allowed to pour in simply because there was no alternative to it. It didn't seem to matter at the time. The gasoline stations on every corner had more gas than they knew what to do with and were selling it for 30 and 40 cents a gallon. Gas wars were common. A fill-up sometimes brought such promotional bonuses as placemats, free steak knives, highball glasses, and a car wash.

Then, to their shock, frustration, and anger, Americans discovered the vulnerability of their supply of petroleum. In October 1973, the Seven Sisters met Arab Big Brother in the oil revolution sparked by the Arab-Israeli conflict. After the outbreak of the Yom Kippur War, the Arab members of OPEC placed an embargo on all oil shipped to the United States and the Netherlands, the major European refining center, because of American and Dutch support of Israel. Two months later, OPEC more than quadrupled the price of its crude oil from $2 to $9 a barrel. The now familiar results were shortages until the embargo was lifted in March 1974; higher prices for petroleum products, which were not lifted but went steadily higher; the sudden awareness that the industrialized world was faced with an energy crisis; and record profits for the oil companies.

The producing governments took a steadily firmer grip on official prices, production levels, and the crude-oil allocation system. Now the major international companies extract about 45 percent of OPEC oil, a sizable amount but only half as much as before. OPEC has steadily shorn the Seven Sisters of most of their power, reducing them almost to the status of contractors, and is making inroads into one of their remaining strongholds—the "downstream" operations of refining and marketing.

But the profits of the Seven Sisters suffered no diminishment. On the contrary, when OPEC initially quadrupled its prices, this action served to quadruple the value of the industry's inventories—stocks on hand. The Seven Sisters, in fact, recovered rather nicely from their confrontation with Arab Big Brother, becoming sort of junior partners in the cartel and providing it with invaluable assistance. As a 1976 report titled *OPEC: History and Prospects of Oligopoly* by the U.S. State Department's Bureau of Intelligence and Research put it:

> Because of their vertical organization, the international oil companies presently are important to OPEC's survival. These companies provide a cartel-like marketing mechanism that allows easy pass-through of crude oil price increases. They also provide crucial expertise to keep national oil companies operating effectively.
>
> Despite the erosion of their positions in producing countries, the international oil companies are unlikely to take any action aimed at breaking the cartel. They are too dependent on OPEC for secure supplies to feed their downstream investment.

In July 1977, the State Department's Bureau of Intelligence and Research made another assessment:

> OPEC relies on the multinational oil companies as a vital link in policing its price-supporting cartel. As long as members use the multinational companies, the cartel guards itself against price erosion. . . . If OPEC members continue to market most of their crude oil through the multinational companies, the threat to the cartel is fairly low.

There is no dark conspiracy at work here. The big oil companies didn't create OPEC or the energy crisis. However, their interests coincide with those of OPEC and are in direct conflict with those of the consumer. Most consumers have a lot to gain from the destruction of OPEC, but the oil companies have no reason to undermine that organization. With OPEC fixing the world price of oil and restricting production as necessary to sustain that price, the industry doesn't have to worry about the intense price competition and falling prices that prevailed before 1973. When the cartel raises prices, the value of all oil rises accordingly, and so do revenues from oil that the companies own and produce elsewhere. When these increases outstrip operating costs, profits soar.

Dr. Avram Kisselgoff, former chief economist of the Allied Chemical Corporation, tried to clarify how price rises by OPEC are converted into higher prices and profits by the oil companies in 1980 Working Paper No. 245 titled *The Propagation of Prices in the Oil Industry, 1958–76* for the National Bureau of Economic Research. In a lucid précis of Dr. Kisselgoff's study in his column "Economic Scene" in the *New York Times* of June 11, 1980, Leonard Silk wrote:

> For instance, refiners' gross margins—the difference between the composite price of a barrel of foreign and domestic crude and the price of a barrel of refined petroleum products—rose from $1.36 in 1972 to $1.78 in 1973 and to $2.72 in 1974. These increases in margins meant additional gross receipts for refiners of about $1.9 billion in 1974.
>
> Dr. Kisselgoff stresses that in the oil industry what counts is the absolute size of the margin and not the percentage markup. The oil industry itself has historically used the concept of margin in absolute terms. While prices rose sharply in 1974 and actual margins increased, margins in percentage terms declined at the refining and marketing levels. With the explosion of crude oil prices and the imposition of stricter price controls on gasoline, the oil companies increased their profits by raising the prices of other products faster than that of gasoline.
>
> Using data from the Chase Manhattan Bank on 29 United States oil companies, Dr. Kisselgoff found that the annual gains

in net income after taxes, which had not exceeded 10 percent in the 1967–72 period, jumped by 70 percent, to $11.7 billion, in 1973 and by 40 percent, to $16.4 billion, in 1974.

The companies held on to a much larger share of that rising income: Retained net income of the 29 companies rose to 65 percent in 1973 and 71 percent in 1974, from 45 percent in 1972. As a result, the net worth of the group of 29 rose by 11 percent in 1973 and 15 percent in 1974.

This rise in net worth helped make the oil companies' profits lower in percentage terms. While their reported profits declined to 12.8 percent in 1975, from 15.5 percent in 1973, their actual net income after taxes remained the same.

Dr. Kisselgoff contends that a "meaningful evaluation" of the profitability of the oil industry should take into account not only the behavior of changes in net worth over time but also many other factors, including the extent to which the oil industry is investing in reserves of fuel other than oil, such as coal, uranium, oil shale, and timberland. As a result, he notes, the oil industry is in possession of assets that do not currently yield profits but "whose potential profit increases inexorably with the rising value of the reserves."

This admirable statement of a complicated issue did not make much of an impression in the frenetic atmosphere of the energy crisis. Since every melodrama requires a villain, many assigned that role to the oil companies in view of their soaring profits—termed "obscene" by Senator Henry Jackson. There was a widespread suspicion that the private companies had created the energy crisis by deliberately holding back supplies to wait for higher prices. That suspicion intensified when, in the aftermath of the failure of the federal government to come up with an effective national energy plan and of the revolution in Iran, a gasoline shortage hit California in April 1979. The shortage soon moved to the East Coast, the price of gasoline went to over $1 a gallon, and—to the bafflement and outrage of many—U.S. oil companies reported huge percentage gains in first-, second-, and third-quarter earnings. The year-end figures were astronomical.

Exxon's profit increase over 1978 was 55 percent, and companies with smaller earnings recorded even more spectacular percentage profit gains. The top spot by that yardstick went to Standard Oil of Ohio, whose 1979 earnings of $1,186,000,000 represented a 163 percent increase over 1978. Texaco came in second with earnings of $1,759,069,000, a jump of 106.4 percent over 1978. Texaco was followed by Mobil, earnings of $2,010,000,000 (78 percent); Gulf, earnings of $1,322,000,000 (68 percent); and Standard Oil of California, earnings of $1,785,000,000 (64 percent). More modest increases were registered by Atlantic Richfield (45 percent on earnings of $1,166,000,000) and Standard Oil of Indiana (which had to be content with a 40 percent jump on its earnings of $1,507,000,000).

According to a CBS-*New York Times* poll taken in the spring of 1979, 69 percent of the public believed that gasoline prices were rising not because there was an energy crisis but because the oil companies wanted higher profits. Suggestions that the American people themselves were to blame for the crisis because they used more oil than they really needed and had neglected to practice conservation got short shrift. As Professor Walter Dean Burnham of the Department of Political Science at MIT put it:

> In a very broad sense, the charge may be valid—in the sense that a whole people is shaped in its patterns of work, life and consciousness by the economic system and the few who control it. Most of us pretty much take life as it is given to us by others. For example, destroy local mass-transit systems, promote suburban sprawl through every governmental and private incentive, permit central cities to deteriorate into jungles and stimulate the automotive industry by every advertising trick known to man, and what do you get? A spread-out network of settlement, work, distribution and consumption which has become absolutely dependent on the automobile for its existence.
>
> Whom do we have to thank for all this? We can be a little more specific than "the American people," can't we? Is our memory so short that we forget the lavish television commercials produced by the oil companies as late as 1973 urging all of us to consume, consume, consume? Do we forget that as late as

the very end of 1973 the President of the United States was saying, "We use 30 percent of all the energy. . . . That isn't bad; that is good. That means that we are the richest, strongest people in the world and that we have the highest standard of living in the world. That is why we need so much energy, and may it always be that way"?*

The hegemonic corporate interests and the political leadership which so often speaks and acts for them have led the country into a frightful dead end since World War II. No doubt they did this with the full consent and acquiescence of the American people. Ultimately they did so because capitalism—whatever else it may be—places every incentive on taking the short view rather than the long, optimizing profits rather than planning, and insuring the supremacy of individual self-interest over the "national interest." The law of least effort has dominated. It has led us to the verge of a gigantic national tragedy.

How easy it is to blame "the American people" for all this when the piper at last has to be paid! For then it becomes unnecessary to think about the implications of the fact that the American dream is about over. The costs of the transition to something else will be agonizingly high. If the political system is not to blow up under the strain, these costs will have to be apportioned with some pretense to equity.

If "the American people" were not to blame, the big oil companies protested that they were not to blame either. After decades of operating in secrecy and disdaining both the media and the public, the companies mounted a huge advertising and public relations campaign to explain their problems. They lamented that they were hamstrung by excessive governmental regulations, and protested that their profits, far from being "obscene," were actually inadequate to finance the job of finding new supplies to counter OPEC.

The industry's newfound candor left many unconvinced. "I don't think an industry could ever have a lower image in the American people's eyes than does the oil industry at this

*Nixon to the Seafarers International Union, Washington, Nov. 26, 1973.

moment," observed Senator Howard Metzenbaum, Democrat of Ohio and a persistent critic of the industry, in the summer of 1979. A year later, at a Senate hearing attended by executives of Exxon, Mobil, and Sun Oil, he said that he felt the oil companies were acting as a "monopoly" to "gouge" consumers and asked: "Why is the oil market in this country so totally exempt from the economic laws that prevail in a truly free market?" Despite a public relations budget in excess of $20 billion a year that was designed to convince the public and its elected representatives that it was responsible and had rational answers to the energy crisis, Mobil found itself labeled by President Carter as "perhaps the most irresponsible company in America."

At the same time the oil executives and lobbyists were going all out to persuade politicians and a mistrustful public that they needed even greater profits for exploration, the biggest oil companies were using their sizable cash flows to diversify into businesses that had nothing whatever to do with exploration or even energy. Mobil bought the Marcor Corporation, which owns Montgomery Ward, the fourth largest retail merchandiser in the United States, and the Container Corporation, the largest producer of paperboard packaging. The price was $1.8 billion, one of the largest acquisitions on record. Exxon began a tender offer of up to $1.17 billion for the Reliance Electric Company of Cleveland. Atlantic Richfield began to buy up Anaconda Copper, and Getty Oil spent $10 million for 85 percent of a television service called the Entertainment and Sports Programming Network and $570 million for the ERC Corporation, an insurance company based in Kansas City, Missouri. Texaco, while placing ads encouraging the conservation of gasoline, moved its corporate headquarters from the Chrysler Building next to Manhattan's Grand Central Station to a new $2 million headquarters in Westchester County. The move not only added to the deterioration of New York's center city, but also obliged some sixteen hundred employees to drive to and from work. Gulf tried for the brass ring with its attempted acquisition of Ringling Brothers–Barnum and Bailey Combined Shows, but eventually backed away from this particular circus.

Such moves, combined with rising prices, shortages, and exploding profits, made oil the industry people love to hate. Some industry executives took solace in the view that the hostility merely expressed a basic human urge to find a scapegoat in the hope that once the villain was exposed, the energy crisis would go away.

Despite all the books, articles, TV shows, and political rhetoric about the energy crisis, Americans still do not have a picture of how the oil business works today and the role it can be expected to play in the future. They, and their elected representatives, know no more about the true nature of the oil business than (as Dad Joiner might have put it) a hog does about Sunday. They know that it is big—and it is, a $360 billion-a-year industry that controls refineries resembling images out of *Star Wars*, supertankers larger than aircraft carriers, offshore drilling rigs taller than the World Trade Center, pipelines longer than any superhighway. The global revenues of Big Oil—Exxon, Mobil, Texaco, Standard Oil of California (Chevron), Gulf, Standard Oil of Indiana (Amoco), Atlantic Richfield (Arco), and the U.S. operations of the British-Dutch–owned Royal Dutch/Shell Group—exceeded $203 billion in 1978.

But there's a lot more to the oil business than Big Oil. Far removed from the boardrooms, banks, and congressional inquiries are the roughnecks, members of a drilling crew; the roustabouts, laborers in a work pool or gang; the derrickmen—also known as derrick skinners, derrick monkeys, skyhookers, and towerbirds—who rack or stack pipe inside the derrick; and the doodlebug men, who use strange devices to prospect for oil-bearing structures. There are the family-owned Mom-and-Pop filling stations in small New England or Sunbelt villages. There are the mysterious brokers who deal for crude oil outside the regularly contracted channels on the Rotterdam spot market. There are hundreds of thousands of elderly pensioners and stockholders whose income derives from oil. And there are more than ten thousand independent wildcatters, who have no connection whatever to Big Oil.

Oil is a diverse and complicated business that touches every facet of American life. It supplies three-quarters of the nation's energy and more than three thousand other products, including the petrochemicals that end up as nylon shirts or plastic bottles and the herbicides, pesticides, and liquid fertilizers upon which the productivity of American agriculture depends. In 1979, the United States imported half of its petroleum and petroleum products, paying some $90 billion for the privilege. Apart from its debilitating effect on the U.S. balance of payments, the dangerousness of this dependence can be illustrated by an example. One-third of America's imported oil passes in tankers through a 2-mile-wide channel in the 18-mile-wide Strait of Hormuz leading from the Persian Gulf into the Arabian Sea. In July 1979, Sheik Ahmad Zaki Yamani, the Saudi Arabian oil minister, commented: "The Palestinians are growing ever more desperate, and I wouldn't be surprised if one day they sank one or two supertankers in the Strait of Hormuz, to force the world to do something about their plight and Israel's obstinacy. This would block the channel through which pass 19 million to 20 million barrels daily. This would make the present crisis seem like child's play."

This kind of vulnerability is not going to be relieved by the discovery of giant new reserves on the mainland United States. The days of Patillo Higgins and Spindletop, Dad Joiner and Daisy Bradford No. 3 have gone the way of the buffalo and the kerosene lamp. In the words of Dr. H. William Menard, director of the U.S. Geological Survey: "The chances of finding a new giant oil field in the United States are just about zero."

Right now, America still has sizable amounts of oil and natural gas, but not enough to keep its economy and military going without imports, and it is running out of these fossil fuels fast. So other energy sources must be developed. The drawback is that in practical and economic terms, an effective transition cannot be made for another twenty years. In the interim, the United States will have to institute serious measures for conserving oil and gas, depend on imports from OPEC and non-OPEC nations like Mexico and perhaps China, and rely on the private petroleum industry for supplies from declining existing wells and new discoveries in remote regions like Alaska, Antarctica, and off-shore areas. The many formidable problems hindering the development of coal and nuclear power must be resolved, and the transition to alternate sources like tar sands, oil shale, solar energy, hydropower, geothermal energy, and windmills must go forward. But none of these sources is going to put a tiger in anyone's tank next year, or the year after.

Chapter Three

The Other Games in Town

Natural gas, a mixture of hydrocarbon gases frequently occurring with petroleum deposits, forms an essential part of the oil game, which is no longer exclusively concerned with oil. Until the early 1950s, natural gas was so plentiful and cheap (it sold for pennies per 1,000 cubic feet) that its producers considered it an unwanted by-product in the extraction of oil and routinely flared—or burned—it away at the wellhead. With oil in scarcer supply, natural gas has been transformed into a premium fuel, clean-burning and relatively easy to transport by pipeline, and 95 percent of American requirements can be domestically produced. At current consumption rates, there is about a fifty-year supply left.

Natural gas is valuable for home heating, electric power generation, and industrial use, especially as a boiler fuel. It represents about one-quarter of the total energy America consumes. There are many independent natural gas producers and pipeline companies, but considering the size of the market and the fact that natural gas now brings about $2.20 per 1,000 cubic feet on the controlled interstate market (controls will be phased out by 1985), it should come as no surprise to students of the integrated oil companies that they control over 70 percent of the reserves and production. The ten leading natural gas producers are: Exxon, Shell, Amoco, Gulf, Phillips, Mobil, Texaco, Union, Atlantic Richfield, and Conoco.

Coal, of course, is America's premier fossil fuel. The nation has enough coal to last more than two hundred years, even at annual consumption rates of 1 billion tons a year. And eleven oil companies now own 25 percent of all the coal in the United States. They are big oil companies, for only they have the cash to buy coal companies. It is not a game for the middle-sized or the small independent.

In 1966, well before the onset of the energy crisis, Conoco (then Continental Oil Company) bought America's second largest coal producer, Consolidation Coal Company. Also in the 1960s, Occidental Petroleum acquired the fifth largest, Island Creek Coal, and Gulf took over Pittsburgh and Midway Coal, whose annual production it hopes to expand to 25 million tons by 1985.

Since 1970, Exxon has become the nation's fourth largest coal reserve owner with over 9 billion tons, which it is developing through two subsidiaries, Carter Oil Company and Monterey Coal Company. Shell owns three coal companies; and ARCO Coal Company, Atlantic Richfield's entry in the coal business, projects annual production of 32 million tons by 1985. Standard Oil of California (Socal) owns 20 percent of AMAX Coal, the third largest producer. Texaco holds approximately 2.3 billion tons of coal reserves at the Lake DeSmet Reservoir in Wyoming. In 1977, Mobil bought Mount Olive and Staunton Coal, and two years later, Sun paid $300 million for Elk River Resources. In 1980, Getty Oil bought the UNC Plateau Mining Company from UNV Resources Inc. for about $60 million in cash. Getty thus acquired a company that owns and operates an underground mine about 100 miles southeast of Salt Lake City that yields about 850,000 tons of low-sulfur coal a year. It also has an estimated 60 million tons of coal in place elsewhere in Utah.

All told, according to the Washington Analysis Corporation, a subsidiary of Bache Halsey Stuart, oil companies account for fourteen of the top twenty holders of domestic coal reserves. A 1980 survey by *Business Week* of the leading twenty-six oil-owned coal companies concluded that oil companies expect to be producing 40 to 50 percent of all the coal burned in the United States by 1985—even without a single takeover of another coal company.

Although alarmists foresee a time when the oil companies will so thoroughly dominate the coal industry that they will be able to restrict competition and supply and to jack up prices, so far the Justice Department has not initiated any antitrust action. Some have even argued that the trend could be a good thing, if the oil companies infuse a stagnant, unstable industry with capital and expertise so that it produces substantially more energy. One expert who disagrees with this optimistic view is S. David Freeman, chairman of the Tennessee Valley Authority, the biggest coal consumer in America. He is trying to reduce the percentage of coal (46 percent) that TVA buys from the oil companies. "We're trying to get in a position so we will not be at the mercy of the Seven Sisters and their fellow travelers," Mr. Freeman has said. "Big oil has the strength to keep fuel supplies short and prices high. They can even keep the sun from rising as an energy competitor."

In view of the possibility that American Big Brother might eventually move to divest the petroleum industry of its nonoil holdings, one wonders why the oil business wants to move into the coal business in such a big way. Superficially, coal does not appear to be so inviting a target. True, the United States is the world's second largest producer of coal (after the Soviet Union) and has perhaps one-quarter of the world's reserves (in the idiom of mining executives, America is the "Saudi Arabia of coal"), but while coal reserves are abundant, they are accompanied by an abundance of headaches.

Coal is a troubled industry prone to slumps with (in 1980) an overcapacity of 100 million tons and twenty thousand jobless miners. Buyers are hesitant to undertake the enormous expense of switching giant boilers from oil or gas to coal. Steel has been in a chronic slump, reducing the demand for high-priced metallurgical coal for coking. Soaring railroad rates have made the cost of moving western coal higher than the cost of mining it.

Environmental problems are formidable and costly. Most

eastern coal is extracted by underground mining, while most western coal can be strip-mined (mined just below the surface). But the federal strip-mining law forbids opening a mine on land that cannot be reclaimed. Environmentalists and legislators in the western states, especially in the Rockies, are not overjoyed at the prospect of having their pristine territories invaded by strip mines, boom towns, and huge demands on scarce water supplies.

Coal is dirty. Mining the stuff underground can cause black lung disease, and burning it in its natural state creates numerous air pollution problems, including the little understood phenomena of radioactive fallout from stack emissions and acid rainfall. The Clean Air Act is designed to restrain such pollution, but it also restrains the full development of America's abundant coal reserves as a substitute for imported oil.

Yet the men who run the oil business seem willing to tackle these problems. Why? For two reasons: anticipated profits and the chance to transform their coal into something other than coal.

As for profits, the electric power industry is the biggest market for coal. Utilities are under increasing government pressure to convert existing oil-burning plants to coal and to burn only coal in new plants. Legislation has been enacted to make available federal funds to help utilities convert from oil to coal. Since nuclear licenses are increasingly hard to get, industrial plants will also have to make greater use of coal. The leaders of the Western alliance decided in Venice in June 1980 to double coal production and consumption by the end of the decade.

When demand picks up, as inevitably it must, the oil companies will be there to supply the coal—in a more concentrated market and at a price that should yield handsome profits later in the 1980s. And as the demand for all kinds of energy becomes more frenetic, there is a good chance that the new conservative White House and Congress will push to abolish the more costly provisions of the Clean Air Act and the strip-mining law. All this is good news for the oil companies, if not for American consumers. An increased demand for coal, if combined with a moratorium on nuclear power, could force up the price of coal so that the utilities that burn it will have to charge their customers more for electricity.

The second reason the oil companies have been so enthusiastically acquiring coal companies is coal's potential value in the manufacture of synthetic fuels. Staring at a future of dwindling crude oil supplies, oil men have concluded that their salvation lies in becoming leaders in synthetic fuel development.

There is nothing new about making oil and gas out of coal. London's nineteenth-century street lamps were powered by coal gas. Hitler's Messerschmitts and Panzers ran on fuel produced from coal. In 1980, South Africa produced 10 percent of its oil and gas from coal, and boasted the only commercially proven oil-from-coal plant in the world.

The drawback has not been technology, but price. Synfuel has always cost more to produce than crude oil. (Coal conversion is economically viable in South Africa because of the use of very cheap black labor in the coal mines.) But with the recognition that the price of foreign crude will not stabilize, domestic synfuel has begun to look more and more attractive. If it can be produced for the same price as the cost of a barrel of foreign crude, say $35, then the oil companies perceive a chance for profit and are prepared to move in this once dormant area.

There are two technologies for transforming coal into oil—liquefaction and gasification—and in both, hydrogen is added to the coal. In the "direct" liquefaction process, coal is dissolved in an organic liquid and exposed to pressurized hydrogen gas. "Indirect" processes heat the coal in the presence of steam and chemically convert the gas to liquid fuels. Exxon, Gulf, Mobil, and Ashland Oil are actively engaged in making liquefaction a commercial proposition.

Unfortunately, both liquefaction and gasification present most of the problems associated with large-scale coal mining. It is also suspected that coal liquids are carcinogenic and that heating coal in the processes will cause pollution.

The oil companies have been active in developing two other alternate energy sources in which they have vast holdings, tar sands and shale. Tar sands are gooey concentrations of tarlike oil locked in surface and shallow underground sand deposits. When the sands are combined with hot water and steam, the resulting mixture can be refined into crude oil and natural gas.

There are some tar sands in Utah, but they're so inaccessible that they would be difficult to exploit. Most of the deposits are located in northwestern Alberta, Canada, and these could yield 26 billion barrels of oil. Cities Service, Gulf, Sun, and the Canadian subsidiaries of Exxon and Shell have experimental projects under way in Alberta, while Texaco Canada has about 575,000 net acres of tar sand leases in the Athabasca area of Alberta. Texaco says that "recovery will depend upon the development of economically viable technology." Translated from oil company jargon, that means when Texaco can turn a profit in recovering the oil.

Shale, once described by the Indians as "the rock that burns," is a hard rock laced with a solid organic material called kerogen. When heated to temperatures as high as 900°F, it breaks down into oil and gas. The world's biggest known deposits of shale are located in Wyoming, Utah, and Colorado. Two processes have been developed to recover oil by "cooking" the shale rock.

Union Oil Company of California has constructed a project 62 miles northeast of Grand Junction, Colorado, in which standard mining techniques bring the rock to a surface retort for heating and converting to oil. Colony Oil, Superior Oil, and a partnership of Gulf and Standard Oil of Indiana are experimenting with a similar technique.

The second, more radical, method is being tested by Occidental Petroleum in Colorado. It involves starting fires in underground mines to separate the oil so that it can be pumped to the surface. This technique avoids the twin problems of how to dispose of huge amounts of spent shale and how to meet the enormous demand for water needed to wash out impurities and for the cooling required by the first method.

Texaco is testing a third process at its shale properties in Utah in which radio-frequency electric fields heat deposits containing immobile heavy hydrocarbons. This process is supposed to produce liquid and gaseous hydrocarbons in place, without the burden of mining, retorting, and waste disposal.

In the spring of 1979, President Carter proposed a ten-year $88 billion effort to construct a network of synthetic fuel plants at which up to 2.5 million barrels a day of crude oil could be derived from coal, shale rock, and tar sands. That would be enough to allow America to cut its projected consumption of imported oil by about one-third by 1990. After bargaining and haggling for a year, the House and Senate agreed to a $20 billion program of federally subsidized synthetic fuel plants, with a production target of 500,000 barrels daily by 1987. That is the equivalent of 8 percent of 1980 imports and 4 percent of all domestic oil consumption.

Not everyone is as enthusiastic about the prospects for shale and tar sands as the oil companies and the federal government. Officials of the western states, local representatives of the Environmental Protection Agency, the Sierra Club, and similar groups are allied in advocating a cautious approach to shale development. The governor of Colorado, Richard Lamm, has said that any crash program "could do irreparable damage to our water supply, to our communities, to our environment."

But Scientists and Engineers for Secure Energy, a committee that includes seven Nobel Prize winners, took a different view. "Our security requires a crash program in energy production, including nuclear energy, oil shale, synthetic oil and gas from coal, and enhanced oil and gas recovery," a report by this scientific group warned in June 1980. This conclusion reflects the "doomsday scenarios" prevalent on Capitol Hill, those estimates of what the impact would be if imports of Persian Gulf oil were sharply cut or stopped altogether. The more draconian scenarios predict losses to the gross national product of billions of dollars a year and an additional ten million Americans unemployed.

The only certain conclusions that can be drawn about synthetic fuels at this point are that they offer no panacea, that they will be very expensive to develop, and that the oil companies will be in the forefront of the development. Even with a crash program backed by tax incentives and government subsidies, the best estimates are that only about 200,000 barrels of shale oil a day can be produced by 1990. That is roughly equal to the amount of crude oil imported from Iran in 1979. The Synthetic Fuels Corporation proposed by President Carter and approved by Congress in June 1980 might reasonably be expected to produce in 1990 the equivalent of about 20 percent of what America's oil production was in 1980, at the staggering cost of more than $70 billion. Synfuels remain one of the costliest, most cumbersome, and least competitive of all energy sources.

The oil companies' nonoil holdings are not limited to coal, shale, and tar sands. Most of them also own substantial uranium deposits. Socal started producing uranium in Texas in 1979, and Texaco has identified some 7 million pounds of in-place uranium resources on its fee and leased holdings in south Texas. Exxon, Mobil, Shell, Arco, Phillips, Conoco, Sohio, Union, Getty, Marathon, and Kerr-McGee own uranium properties from Wyoming to Australia.

The oil companies have rationalized their heavy diversification into energy sources other than oil and gas on the grounds that it will provide America with more needed energy and stave off the "doomsday scenarios." There is, of course, nothing illegal about diversification, which is a common practice in all industries with large amounts of cash on hand. The men who run the oil business are, after all, responsible to their boards of directors and stockholders for showing a profit, and diversification is one route to that goal. In a democratic free enterprise system, no one could object to an industry's spending its earnings in whatever productive way it saw fit, and the oil business must guard against the day when the oil runs out. Nevertheless, oil differs from other businesses in one crucial aspect—it is a private business whose operations have a tremendous impact on the public good.

It is the extent of the oil companies' diversification into fields that have little or nothing to do with providing energy that is startling. They are really not traditional oil companies anymore, but could better be described as "energy conglomerates" or "natural resource managers." Although their p.r. campaigns would have Americans believe that they are spending their swollen profits on a heroic quest for more energy, this comforting assurance is not wholly true. The companies are, in fact, using quite a bit of their recent profits to buy up businesses that have nothing to do with energy. The justification they offer is that some nonenergy diversification is necessary to offset the depletion of crude oil and that profits derived from other areas will be plowed back into finding oil in hostile regions. This claim has some merit, although there are those who strongly disagree with it.

Senator Howard Metzenbaum, Democrat of Ohio, introduced a bill along with Senator Edward Kennedy and eleven other senators to impose a ten-year moratorium on acquisitions by sixteen oil companies of other giant companies (it was not enacted). In a letter to the *New York Times* dated October 30, 1979, Senator Metzenbaum, responding to a Mobil advertisement that criticized his position, wrote:

> The point of the bill is simple: The national interest requires the oil companies to use their enormous profits to produce and develop oil and energy. And those profits are going to swell even more in future years. Decontrol of oil prices will take from the American consumer and hand over to the companies as much as $160 billion in the next decade, even with a windfall profits tax.
>
> These vast sums should be used to produce more energy and develop new technologies. They should not be used to buy up insurance companies, retailers, box manufacturers, truckers, hospital-supply, office-equipment, electric motor companies and the like, as the oil companies have done. Gulf even considered buying Ringling Brothers Circus.
>
> As usual, the industry's tactic is to accuse its critics of misrepresentation. Let's see who's misrepresenting what.
>
> • Mobil says decontrol is no special favor because other industries are not controlled. But only the oil industry has been the beneficiary of tightly restricted supply by means of import quotas, production controls and now foreign cartels. These restrictions, incidentally, kept American oil prices well above foreign prices for many years.

• Mobil says non-energy acquisitions were only 2 percent in 1977. But there is a trend, as many observers have noted, that will get stronger with the added cash billions in the next 10 years. In the first four months of 1979, for example, the oil majors proposed to spend over $2.6 billion on 21 acquisitions.

• Mobil says it spent $86 million on research and development in 1978, no "paltry sum." But as a percentage of its $1.2 billion after-tax profits, it is a very small fraction of what many other industries regularly spend on research and development.

• Mobil says it's not "fair play" to single out the oil industry in this legislation. But no other industry will get such huge unearned profits. And the oil companies have other special privileges, which have made them into what one investment analyst called "great cash-flow machines, expected to flow into the early 1980's on a sea of cash." What other industry so totally controls the supply of such a vitally needed product?

The oil industry is trying to have it both ways: It attempts to justify its resistance to a windfall profits tax on its excess profits arising from decontrol in the name of energy development, but then insists on its right to use these profits to buy up large businesses not related to energy.

While the average American is wondering how to pay for huge increases in heating oil and gasoline prices, the oil industry is wondering what to do with all its profits.

Mobil headlined its advertisement, "Senator Metzenbaum ought to know better."

Mobil *does* know better.

Whether the senator or Mobil knew better just how much Mobil and other oil companies were involved in ventures having nothing to do with oil is an interesting question.

The integrated majors have been active in chemicals in a big way for years. For Exxon, Gulf, Shell, and Standard Oil of California, chemicals represent a multibillion-dollar investment. Texaco spent $500 million on its new petrochemical plant at Port Arthur, Texas, which became operational in 1978. The company's worldwide net earnings from petrochemical operations amounted to $96.6 million in 1979.

Chemicals would seem to be a justifiable business for the

petroleum industry to be in, since the chemicals it produces are petrochemicals refined from petroleum hydrocarbons. Among the hundreds of petrochemicals are ethylene oxide and ethanolamines, used, respectively, in the manufacture of detergents and antifreeze; olefins, used mostly in the manufacture of end products by the cosmetics, plastics, apparel, and rubber industries; and insecticides, herbicides, and pesticides. The outside industries, particularly agriculture, that depend on these petrochemicals would be hard pressed if their supply were cut off, but it might be noted that chemicals contribute nothing to the energy supply and that the oil companies are diverting petroleum into refining and producing them.

Metals and minerals have provided an attractive outlet for Big Oil's embarrassment of riches. Getty, Cities Service, Standard Oil of Indiana, Pennzoil, and Standard Oil of California own copper-producing properties, while many others own various nonferrous deposits like molybdenum. In fact, Socal had acquired 20 percent of Amax Inc., the nation's leading supplier of molybdenum, in 1976 but was rebuffed in March 1981 when it offered to take over the remaining 80 percent for nearly $4 billion, the largest takeover bid on record at that time. Shell produced and sold $1.2 billion worth of copper, zinc, aluminum, and nickel in 1977. Exxon owns a Chilean copper producer and in June 1980 consolidated all its mineral operations, involving such materials as copper, zinc, lead, and uranium, into one unit, Exxon Minerals Inc.

Upon his appointment as president of the new unit, William McCardell provided an insight into why Exxon was expanding its mineral operations. "I will be reviewing various milestones of important projects as they come forward," Mr. McCardell said. "This is a very small part of Exxon at this time. Our objective over the decade is to reach a point where it will be a substantial contributor to Exxon's profits."

Atlantic Richfield bought the Anaconda Copper Company in 1977. "We think of ourselves as a natural resources company," observed Robert Wycoff, Arco's senior vice president of finance and planning. "Anaconda is a mining company and mining isn't that different than oil production." However, this was one nonenergy acquisition upon which Big Brother—in this case the Federal Trade Commission—blew the whistle. It brought antitrust action against Arco, alleging that its purchase of Anaconda would lessen competition in the copper and uranium industries. In June 1980, Arco agreed to divest itself of an estimated $473 million to $573 million of copper interests and promised not to acquire other copper properties for from five to ten years after the effective date of the settlement.

Getting still further afield from the oil business, Getty, Tenneco, and Kerr-McGee own timberland, while Sun, Union Oil of California, and Mobil own condominiums (Mobil's is in Hong Kong). Standard Oil of California owns 65,000 acres of agricultural land in the San Joaquin Valley, while Tenneco has cattle ranches in California and Arizona. Arco converted five hundred of its blue-and-white gasoline stations into convenience food stores called AM/PM markets in 1980.

Exxon introduced Qyx, a computer-programmed typewriter designed to undersell word-processing Xerox and IBM machines, in 1978. Gulf acquired the new town of Reston, Virginia, and a chain of trailer parks, but liquidated most of these holdings when they proved unprofitable. And, of course, in the biggest acquisition of all Mobil spent $1.8 billion to buy Montgomery Ward and the Container Corporation of America. Sun Company (formerly Sun Oil Company) owns Totem Ocean Trailer Express Inc., a West Coast sea transport firm; three convenience-store businesses, Stop N Go Foods Inc., Fast Fare Inc., and Mr. Zip Inc.; and Sun Ship, Inc., of Chester, Pennsylvania, one of the nation's twelve major shipyards. However, the Sun Company phased out all new-ship construction in early 1981, although it planned to continue to repair ships and make industrial products at Chester. The three largest shareholders in the Cetus Corporation, an advanced biotechnology and genetic engineering company, are Standard Oil of California, Standard Oil of Indiana, and the National Distillers and Chemical Corporation, who own about 50 percent of Cetus's shares.

Big Brother has not yet enacted legislation to make the oil

companies get out of nonenergy fields. Industry spokesmen contend that if they were barred by the government from all diversification, they would one day be forced into liquidation, given the fact that the world will eventually run out of oil, and that they are still spending most of their money on finding and developing new energy supplies. This view is given some weight by available facts.

According to Treasury Department data, investment in unrelated fields dropped from 10 percent of cash flow, or $4 billion, in 1974 to 2 percent, or $626 million, in 1977. The five American sisters, Exxon, Mobil, Texaco, Socal, and Gulf, put two-thirds of their worldwide capital and exploration expenditures into the exploration and production of oil and gas in 1979, roughly 15 percent into refining and marketing, 10 percent into other forms of energy and nonenergy operations, and 5 percent into chemical operations. The *Oil & Gas Journal,* a trade publication, says that the entire industry allocated $50.5 billion for capital projects in the United States in 1980, and $10.5 billion for foreign spending.

Sums of this size are hard to get a handle on. But the oil game has always been a money game in which individuals and companies have made, lost, and spent huge sums. A closer look at this money game, whose 1979 profits were smoking-gun proof to many that the companies were ripping them off on energy prices, reveals some surprises.

Chapter Four

The Money Game

Our analysis of the oil game begins with the undeniable fact that it involves very large sums of money and big money has been made by Big Oil. Originally it was made by easterners like the Rockefellers (Standard Oil), Mellons (Gulf), and Pews (Sun Oil) who, in the best manner of the free enterprise system, invested their capital in a fledgling industry whose production was largely located in the Southwest and California. As Texas gradually became a major factor in the oil business, some native sons got into the game without eastern backing or ties to the major companies and joined the ranks of the superrich. Some of these legendary, and often eccentric, characters—like Howard Hughes, whose father invented what is still the premier drill bit, Hugh Roy Cullen, "Bob" Smith, Sid Richardson, Clint Murchison, Sr., and H. L. Hunt—and their descendants became very rich indeed.

Hugh Roy Cullen, who died in 1957, may have been the richest oil man in Texas. He was born on a farm near Dallas in 1881, left school in the fifth grade, sacked candy, toiled as a cotton buyer, and entered the oil business in 1917, leasing properties in central Texas that turned out to be gushers. His independent operation, the Quintana Petroleum Corporation, is a Houston-based production company that manages properties and sells the oil and gas to major companies for refining. This is the usual practice for independents and explains why Quintana is not a household name like Exxon or Gulf.

When a biography of Cullen was published in 1954, he spent more than $216,000 to have 108,000 copies of the admiring volume sent to individuals, newspapers, and libraries. Cullen gave away a lot of the money he made; his gifts to the University of Houston alone came to more than $50 million. One of his five children, the late Lillie Cranz Di Portanova, married into the Italian nobility. Her sons, Baron Enrico Di Portanova and his brother Ugo, received $13 million in 1979 in gross income from a trust established by Cullen in 1955.

Sid Richardson never finished college. He made his first big strike in the Keystone field in west Texas while in his early forties. By the 1940s, his fortune was estimated at more than $150 million. A bachelor who never carried any money with him, Richardson died in 1959; John Connally, who handled his legal affairs, was named co-executor of his estate for a fee of $750,000.

The late Clint Murchison, Sr., who came from the same small Texas town as Richardson, did equally well. Today, his sons, John D., who went to Yale, and Clint W. Jr. (Duke, MIT), have diversified their father's fortune into insurance, banking, aircraft, minerals, publishing, venture capital, and sports (Dallas Cowboys), holding an interest in at least one hundred companies. The Murchison brothers are part owners or directors of enterprises worth more than $1 billion.

For connoisseurs of Texas gothic, however, the saga of the Hunt family illustrates better than any other how much money has been made in the oil game by independents with no ties to what is commonly thought of as Big Oil. The extent of the vast and hidden Hunt empire only came to light in 1980 when two brothers, Nelson Bunker Hunt, then fifty-three, and William Herbert, fifty, took a bath in silver speculation. The Hunt holdings are entirely private, and the family has a mania for secrecy, but the silver debacle forced the brothers to disclose their personal holdings to banks to get a $1.1 billion credit agreement they needed to pay off their debts. At the same time, they were subjected to the indignity of having their silver machinations scrutinized by two congressional subcommittees.

The improbable saga begins with H. L. Hunt, who parlayed a $5,000 inheritance into an estate said to be worth $2 billion in

1954. H.L., who promoted far-right political causes and crawling as an exercise, died in 1974. The patriarch left his fortune, which had grown to an estimated $5 billion by 1979, in trust to his two families. The so-called first family consists of the six living children from H.L.'s first marriage, among them Bunker and Herbert, as they are known, and Lamar. The six received, among other assets, shares in the Placid Oil Company. The remainder of the estate—most of the original Hunt Oil Company—was left to the "second family": H. L.'s second wife, who still lives in the Hunt mansion in Dallas, which is modeled after Mount Vernon, and her four children. The two families' interests are managed separately and sometimes they feud with each other.

Apparently, there was also a third family. In 1978, one Frania Tye of Shreveport, Louisiana, claimed that she, too, had once been married to H. L. Hunt, and that her four children, two of whom were still living, were his. She sued for a share of the estate for herself and her two children. Early in 1979, the first two families, in an unusual display of harmony, agreed to settle $7.5 million on the third family.

The best known Hunts are Herbert and Bunker. Not that the brothers seek publicity. They drive their own cars, are listed in the Dallas phone book, and shun Dallas high society. Herbert lives in a middle-class neighborhood and collects rare coins and U.S. silver dollars. Bunker neither drinks nor smokes, weighs 300 pounds, favors inexpensive chocolate-brown suits, and is on the national council of the John Birch Society. Among numerous other benefactions to Christian causes, he has pledged $10 million toward a $1 billion fund-raising effort by the Campus Crusade for Christ and spent $3 million in 1978 bankrolling the Warner Bros. film *Jesus*.

Using their oil holdings as the foundation, the Hunt brothers have led their branch of the family into other profitable ventures. It owns, among other things, millions of acres of ranches and farmland, including 4 million acres in Australia alone; America's largest producer of beet sugar; and four hundred Shakey's Pizza Parlors. Bunker has been notably successful at horse breeding and owns the largest string of thoroughbred racehorses held by an American; about six hundred of them wear his light and dark green racing silks. In 1976, he became the first American to win both the English and French derbies. One horse he bought for $25,000, Exceller, later earned $1.5 million in prize money. The third brother, Lamar, has become a successful professional sports entrepreneur. He owns the Kansas City Chiefs football team, and an interest in the Dallas Tornado soccer team, the Chicago Bulls basketball team, and World Championship Tennis.

After Libya nationalized his 8-million-acre oil holdings in 1973, Bunker ventured into silver in cosmic fashion. The price was then around $3 an ounce. Bunker made the comment: "Silver looked safer than overseas oil concessions the way things were going. And precious metals were a good hedge against paper money."

Investors like the Hunts routinely buy commodities largely on credit, or margin, which means they make a down payment of a small percentage of the total purchase price. They then own a futures contract, which requires delivery of the commodity at a specified price on a specified future date. The value of a contract increases or decreases depending on whether the price of the commodity rises or falls during the period of the contract. If it falls, those who have lent money to the purchaser make margin calls for additional collateral.

The value of the Hunts' futures contracts had risen nicely by January 1980, when they managed to corral almost two-thirds of the world's privately held supply of silver and the price climbed to $52.20 an ounce. *Time* magazine (January 14, 1980) reported:

> As the nation's paramount silver hoarder, Hunt and at least one brother, William Herbert Hunt, have amassed a new Comstock Lode. Over the past nine months they have earned an estimated $2 billion to $4 billion, and one former business associate sets the Hunts' silver holdings at 100 million oz. Even for a man who could play monopoly with real money, the silver boom is stunning.

As it turned out, it was the Hunts who were stunned. The price of silver began to fall, and by March 27, it had plummeted to

$10.80 an ounce. The Hunts were caught in a classic squeeze. If they sold off their silver to raise the cash needed to meet their margin calls, they faced the near certainty of pushing down the price still further and touching off a snowballing process. Wall Street sources estimated that the Hunts lost more than $1 billion in just a few weeks in March.

There was nothing else for it but to go to banks for a $1.1 billion line of credit to refinance their silver debts. But the group of major banks in the United States and Canada that made the loan laid claim to nearly all the brothers' assets, so the Hunts had to mortgage or sell nearly everything they owned to pay off the loan and its interest of more than half a million dollars a day.

The Hunts had always displayed a passion for avoiding the limelight and an almost pathological fear that their associates would take advantage of them. In 1975, they were acquitted on charges of illegally tapping the telephones of former employees they were convinced were stealing from one of their father's companies; private detectives in their employ were convicted. Now, because of mortgage disclosures made in more than three dozen courthouses across the country, the extent of the privately held empire of the Hunt first family became public knowledge (the other families were not involved in the silver debacle).

The mortgage papers disclose that Bunker had to mortgage, among other personal items, his flock of Boehm bird statues and his cotton plantations in Mississippi and Louisiana. Herbert went in hock for his Greek and Roman statues and twenty bags of U.S. silver dollars (each bag worth $1,000). Lamar, who was not active in the silver speculation, mortgaged his wife's blond mink coat, his Mercedes-Benz, and his Rolex watch. But it was the disclosures about the first family's holdings in oil and other energy ventures that offered a rare glimpse of how an independent, privately held empire operates in the oil game. The centerpiece of the family energy holdings is the Placid Oil Company, owned by trusts established for the six children.

The Texas Railroad Commission said that Placid produced 509,000 barrels of oil for the year ended April 1979, ranking it No. 100 among about seven thousand producers in the state. In

Louisiana, Placid is involved in the Black Lake field, one of the state's largest, and has substantial holdings in the Gulf of Mexico, Utah, Wyoming, and the Canadian province of Alberta. The company owns 9.2 percent of the Louisiana Land and Exploration Company, which had 1979 revenues of $817 million, and 11.3 percent of Gulf Resources, a Houston energy and natural gas resources company with $510 million in sales in 1979.

In addition to Placid, the first family owns Penrod, the largest privately held drilling contractor, with one hundred rigs; some 2.5 million tons of coal reserves; and interests in various gas, sulfur, and uranium deposits. Bunker Hunt personally owns more than $1 billion worth of drilling rights to potential oil-bearing property in Canada's Beaufort Sea, 20 percent of which he pledged to the Engelhard Minerals and Chemicals Corporation to cover debts he incurred when silver prices collapsed.

The money derived from these oil and other energy holdings is so great that the first family empire is unlikely to collapse as a result of the silver debacle, although there was some negative fallout. The banks have barred Bunker and Herbert from commodity and stock speculation until the loan expires in 1990. And with most of their assets in hock, the brothers won't be able to borrow to finance future business deals of any size. However, as long as the oil is there, the Hunts will survive. Sim Trotter, an investment analyst for the Dallas firm of Rauscher Pierce Refsnes, summed up the feelings of that city's financial community when he said that the silver losses of Bunker and Herbert amounted to "a burr under their saddle, but I don't think it's anything that would jeopardize the overall Hunt financial basis."

Congress introduced tough legislation to change the rules for silver trading. In a free enterprise system, of course, high rollers like the Hunts cannot be compelled to plow back their vast oil-derived wealth into ventures that would increase the present oil supply instead of speculating in volatile commodities like silver.

Producing wells like those of the Placid Oil Company provide the principal source of profit in the oil game, but big money can be made in more arcane areas, as the career of Harrell Edmund (Eddie) Chiles of Fort Worth attests. Seventy years old in 1980, Chiles worked as an oil field roustabout as a Texas youth before getting a degree in petroleum engineering from the University of Oklahoma. Then he pioneered the method of acidizing wells to increase production. In 1939, he helped found the Western Company of North America with two oil-well service trucks. Forty-one years later, Western owned thousands of trucks and several hundred million dollars' worth of offshore drilling rigs.

Western's business isn't glamorous, but it's very profitable. It provides four services: offshore drilling; acidizing, which etches channels through the bedrock in which mainland oil and gas have been found so that they can flow freely; cementing, which bonds the well pipe to the side of the drilling hole; and fracturing, or using pressurized fluids to force open the bedrock.

In 1975, Western's offshore drilling business netted less than $200,000. Five years later its annual earnings were estimated at $14 million. Small wonder with an offshore drilling boom on and contracts for jack-up rigs, which drill in shallow Gulf of Mexico waters, ranging up to $33,000 a day in 1980 compared to $15,000 to $20,000 the previous year.

The times have been good to the other three basic services Western offers for obvious reasons. In 1978, price-controlled new oil in Texas sold for about $13 a barrel. Two years later, with price controls removed, it drew as much as $40 a barrel and the attention of hundreds of wildcatters, who rushed into fields previously disdained as marginal. Any field turned into a producer needed cementing, fracturing, and acidizing. The former roustabout and his family owned Western Company stock worth $160 million in 1980.

Eddie Chiles, who became principal owner of the Texas Rangers baseball team in 1980, would have remained as obscure as the Hunt brothers were before their silver ventures had he not become intrigued by a line from the film *Network* (which he never actually saw). In *Network,* the newscaster played by Peter Finch cries: "I'm mad as hell, and I'm not going to take it anymore!"

Chiles was mad, too, at what he perceived to be the tyrannical bureaucracy, runaway regulation, and wild spending of Big Government. In 1977, he began sponsoring his "What's Wrong with America" sermons on radio. Three years later, they were playing twice daily on 465 radio stations in fourteen western and midwestern states. A typical segment went like this:

> ANNOUNCER: Are you mad today, Eddie Chiles?
> EDDIE CHILES: Yes, I'm mad. I'm sad for the Americans who are trying to raise a family and trying to buy a home when the liberals in Washington are spending more and more to destroy the American dream. You get mad, too.

In addition to the radio broadcasts, Chiles had stickers printed reading: "I'm mad, too, Eddie." Over 200,000 of them were put on hard hats and bumpers from Houston to Denver. Soon Mad Eddie became a celebrity in the Southwestern Oil Patch and the Midwest and planned to put his oil money where his mouth was by expanding his radio schedule and trying a TV campaign. In his broadsides, Mad Eddie never says he's mad about the rising oil prices that benefit his firm or that Western writes the broadcasts off as a business expense (they carry a plug for the company) or that he has taken almost $108 million in federal loan guarantees to build offshore drilling rigs.

You don't have to be a Texan or an antiliberal to profit from the oil game. About the only thing Senator Edward Kennedy has in common with Bunker Hunt and Mad Eddie Chiles is that he, too, profits from independent oil operations.

In the late 1950s, after he had made most of his fortune, Joseph P. Kennedy established Mokeen Oil, with headquarters in Corpus Christi, and the Kenoil Company, incorporated in Delaware. These companies handle the oil royalty properties and leases in Texas, New Mexico, Kansas, and Louisiana in which Joseph P. Kennedy invested. The exact dimensions and holdings of the Kennedy family fortune in real estate, securities, and oil are concealed by a curtain of privacy as forbidding as that rung down by the Hunts. But the family's oil holdings alone are said to be worth many millions—and their value rises with the price of oil.

When Senator Kennedy became chairman of the Senate Judiciary Committee in 1978, he placed the assets he had inherited from his father, among them his oil holdings, in a blind trust administered by others to avoid the appearance of conflict of interest. In disclosing his assets, including stock in Texaco and Standard Oil of California, the senator reported an interest in 103 royalty properties, from which payments are received from drilling and from developers to whom mineral rights are leased. He also reported an interest in thirty-nine oil leases, rights to which are leased to developers while a one-eighth royalty interest on any oil production is retained. Senator Kennedy put a combined worth of $881,000 on these holdings.

In an irony that Bunker Hunt and Mad Eddie Chiles might not appreciate, Senator Kennedy has been a persistent critic of the tax loopholes enjoyed by the oil industry while personally profiting from those same loopholes. The senator and his family have benefited over the years from deductions for the intangible costs for drilling for oil and gas, the depletion allowance, and other tax shelters. And yet, as a senator, Kennedy stirred the ire of oil men by leading the successful fight in 1975 to strip the major producers of their 22 percent depletion allowance (it was retained by the smaller firms such as the ones in which the senator has holdings), by voting to tax oil companies on windfall profits, and by strongly opposing the decontrol of oil and gas prices.

In 1978, according to production records of the Texas Railroad Commission, the Mokeen Oil Company operated eight active natural gas wells that produced 540,922 cubic feet of gas, and seven active oil wells that produced 11,628 barrels of crude —respectable figures, but hardly huge by Texas standards. In that same year, Mokeen was charged by the San Antonio enforcement office of the Department of Energy with "improperly" claiming an exemption on price ceilings on oil and natural gas. In January 1979, Mokeen signed a consent order with the department in which it neither admitted nor denied guilt but agreed to reimburse producers for the alleged overcharges of $72,000.

Supporters of the senator point out that his oil holdings are in a blind trust that he is not allowed to know anything about. They maintain that his positions on price decontrol and the oil industry prove that his public policy positions are not influenced by his personal finances. Some have suggested that the irony would disappear if those who manage his finances would just get out of the oil game.

The Belridge heirs, whom few people had ever heard of before the fall of 1979, did just that and reaped a bonanza that might have given Joseph P. Kennedy or H. L. Hunt pause. This story begins in 1911, a long way from Massachusetts or Texas. In that year, a fruit grower, Frank Buck, Sr.; a financier, Burton Green; oil man Mericos (Max) Whittier; and two real-estate men bought some land in the San Joaquin Valley 65 miles west of Bakersfield, California, that has been described as something "a coyote would have turned up its nose at." The purchase was primarily a land speculation, although Whittier had noticed oil seeping from the ground. The Belridge Oil Company was formed. When the real-estate men sold out to Mobil and Texaco in the 1930s, members of the Whittier, Buck, and Green families held on to 56 percent of the company's outstanding shares.

It did not seem like a momentous decision at the time. Belridge's oil production was tiny through the mid-1960s, its stock was selling for only $60 a share, and its name wasn't even on its modest headquarters on the fringes of Los Angeles. The problem was that its oil was heavy oil, higher in density than the conventional or light oil on which the world normally runs. Heavy oil is difficult and expensive to extract.

Then improved technology made it possible to heat heavy oil so that it flows to the surface. Even so, Belridge's production had only reached a modest 16,000 barrels a day by 1974 because the steam technique made it three times more costly than conventional oil. But in that year, price controls were removed from the low-yielding "stripper" wells that account for 90 percent of Belridge's production.

The game changed immediately when Belridge could sell its crude at market rather than controlled prices. From revenues of

$69.3 million in 1974, the company reported a rise by 1978 to revenues of $155.8 million and profits of $43.9 million, more than double 1974 income. Dividends jumped from $5 a share in 1974 to $21 in 1978 for the 550 shareholders.

This sort of action attracted the attention of the major oil companies, always on the lookout for secure sources of profitable domestic crude and with plenty of cash on hand for acquisitions. They noted that Belridge had 376 million barrels of proven oil reserves on 33,000 acres of land in Kern County.

In August 1979, the Whittier, Green, and Buck heirs decided to sell the company as a unit at an auction supervised by the investment banking firm of Morgan Stanley and Company. As majority shareholders, they had the right to do this under California law without obtaining the permission of the minority shareholders.

More than a dozen major oil companies began a spirited bidding on September 17, 1979, including Veba A.G. of West Germany. By the beginning of October, some over-the-counter traders were quoting the issue at $3,000 bid (it had sold for around $470 a share in late March). The auction winner was Shell Oil, whose bid of $3.65 billion was the costliest corporate buy-out ever. (In 1924, Shell had had the chance to buy the company for $8 million, but turned it down.) Shell obviously calculated that it could make money out of Belridge, even after paying out $3.65 billion.

Shell's victory made Mobil and Texaco, which together held 35 percent of the stock, unhappy. They sued to block the sale, contending that it would jeopardize their minority interest. Belridge countered that the two companies had access to inside information through their "agents" on the board of directors, and accused Mobil and Texaco of attempting to "extort from Belridge a preferential distribution of Belridge's oil-producing assets as the ransom for giving up their opposition to the auction program." Shell and the heirs were vindicated when a federal court judge upheld the sale.

So big money has been made by individuals in the oil game, although the number of those who tried and failed would fill the

Houston Astrodome. But the days of Hugh Roy Cullen, H. L. Hunt, and other individual oil barons have vanished as surely as the supply of cheap, plentiful energy. The old barons were all independents, accountable to no stockholders or boards of directors, with no connections to the major oil companies except to sell them their crude. They bought up leases and found oil on them, a lot of it, under the American system of private ownership of subsoil mineral rights in an era of the depletion allowance and negligible taxes.

Now there are no more major oil fields to be found in the mainland United States. The federal government owns offshore rights and auctions them off to the big companies, who alone have the cash to bid for them. Today no canny individual could become as wealthy as a Saudi Arabian prince at the expense of the American consumer. True, the heirs of the old entrepreneurs are bound to make still more money as all price controls are removed from crude oil. One can be outraged at (or envious of, depending on one's point of view) this development, but there is a difference between earning legal profits, however enormous, and conspiring to profiteer at the public's expense. And not all royalty owners are in a class with the Hunts. According to Senator David Boren of Oklahoma, a 1980 study of Oklahoma royalty owners found that 54 percent received a check of less than $200 a month.

Not that serious money can't still be made—and lost—by relative newcomers to the oil game, as witness the experience of the freewheeling Raymond K. Mason, chairman and chief executive officer of the Charter Company of Jacksonville, Florida.

Charter held a modest array of land-developing, banking, and mortgage companies in 1968 but no oil interests when Mason decided to buy sixty small gas stations in the Southeast. A year later, he acquired a small interest in the Iranian oil consortium. He also bought up communications and insurance properties, including *Ladies' Home Journal*, *Redbook*, and *Sport* magazines, six radio stations, Downe Communications, and Louisiana and Southern Life Insurance. In 1970, he bought a small refinery in Houston and $70 million worth of Signal Oil and Gas properties.

In 1974, as a result of the price rises touched off by the Arab embargo, industry earnings soared and so did Charter's, to $39.6 million. The next year earnings plunged to $5.4 million as oil fields the company had acquired in Venezuela were nationalized, the value of its real estate holdings dropped, and the magazine division turned flat. Mason began looking around for more oil acquisitions.

Enter a man with serious problems: Edward Carey, the older brother of Governor Hugh Carey of New York. Edward Carey was the sole owner of Carey Energy Corporation, one of the largest privately held oil companies in the world. He owned Nepco, the New England Petroleum Company, which distributed heavy fuel oil to factories and utilities up and down the East Coast, and a 65 percent share in a Bahamas refinery, the fifth largest in the world (the other 35 percent was owned by Standard Oil of California).

Carey's once lucrative corporation was going broke both because it was totally dependent on foreign crude and because of a quirk in federal regulations. Under the entitlements program, crude oil costs were equalized throughout the U.S. refining industry. But because Carey's refinery was in the Bahamas, it was excluded from the program and had to pay several dollars a barrel more for crude oil than its competitors did. By 1978, Carey was losing $10 million a month and was nearly bankrupt.

Mason stepped in where others had good reason to fear to tread. He offered to buy Carey Energy. Not too much money was involved—$4 million in actual cash and $26 million in convertible preferred stock. Carey, who was faced with receivership and owed $550 million to his suppliers in Libya and Iran, had little choice but to go along with the deal. It was concluded on May 1, 1979.

Then things started to happen, due partly to timing, partly to luck, and partly to Mason's willingness to gamble. The Department of Energy revised its entitlements program to make the Bahamas refinery competitive. The Libyans and Iranians, who had declined to deal any longer with Carey, were persuaded by Mason to resume supplying the refinery at prices that included a premium, and he began paying off Carey Energy debts. The world oil market became tighter than it had been since the embargo, and prices and refining profits soared. Charter's investment of $30 million was turned into a $300 million profit in 1979 after only eight and a half months.

In 1980, the fifty-three-year old Raymond K. Mason could sit in his elegant boathouse office on the St. John's River, which slices through Jacksonville, Florida, and survey Charter's performance the previous year with some satisfaction. Profits had increased 1,468 percent over 1978, to $365 million on revenues of $4.3 billion. Earnings per share went from $1.17 to $14.83. Charter, now the twentieth largest U.S. oil company, had just concluded a deal to get up to 150,000 barrels of crude a day from Alaska, was experimenting with a promising underground shale-oil recovery technique, and was looking for a producing company to add to its crude supplies.

Nearly 90 percent of Charter's revenues and almost all of its profits came from oil, although Mason did not neglect the other two parts of his troika—insurance and communications. He used the oil money to buy Crum and Forster Life Insurance and 50 percent of the nation's thirteenth largest, but failing, newspaper, *The Philadelphia Bulletin*. Edward Carey had no cause to regret the intervention of Raymond Mason into his tangled petroleum activities. Carey got not only the $4 million in cash, but also a $200,000 a year consulting job with Charter and saw his convertible preferred stock become worth more than $100 million.

Mason himself, however, had cause to regret that his company was concentrated mainly in the refining and distribution end of the oil game. Without its own production, that meant that it had to pay sky-high prices for crude, some of it imported from Libya at a fearsome price of $37.50 a barrel. Combined with weak margins for products caused by the oversupply that existed in 1980, the squeeze resulted in earnings for the year of only $50.2 million, a drop of 82.3 percent from the $365.3 million Charter earned in 1979. But Mason should be able to turn this climate of depressed earnings around, now that all price controls have been removed from domestic crude, all refiners pay the same world price, and motorists pay what the market will bear at the pump.

On February 20, 1981, Charter announced that it had hired former president Gerald Ford as a consultant. A company spokesman said that one of Mr. Ford's first duties would be to accompany Charter officials abroad to secure long- and short-term crude oil supplies and to confirm certain credit arrangements in Europe. From Raymond Mason's viewpoint one can only hope that the former president will be more successful on Charter's behalf than was Billy Carter, who apparently once thought that he, too, could profit from the oil game.

In April 1979, the president's brother sought out Jack McGregor, whom he had met in the Marine Corps and who was then executive vice president of Carey Energy. Billy Carter told his old buddy that he was doing some work for the Libyans and could get some Libyan crude oil for Carey Energy. Since Carey was being taken over by Charter, McGregor referred Billy to the Charter Crude Oil Company, which was owned by the Charter Oil Company, which was owned by the Charter conglomerate.

Charter Crude was interested in Billy's suggestion for two reasons. First, Libyan crude is a high-quality, low-sulfur oil and is cheaper to transport by tanker to the United States than oil from the Persian Gulf. Second, smaller independents like Charter do not have the crude oil resources of the majors, and to stay alive economically, they must get it for their refineries from wherever they can. Charter was buying crude from Alaska, Texas, Mexico, Abu Dhabi, and Saudi Arabia, and about 100,000 barrels a day from Libya itself on August 17, 1979. On that day, Lewis Nasife, president of Charter Crude Oil, called on Billy Carter at his home in Buena Vista, Georgia, in a meeting set up by Jack McGregor. The president's brother suggested that he could secure an additional 100,000 barrels a day for Charter Crude and wondered how much commission the company might pay. Billy was now entering the tortuous—but not illegal—world of the petroleum broker.

Large exporters like Libya sell their crude in three principal ways. The first is by making long-term deals with private oil companies, which produce the oil and turn over a portion of the total production to the host nation's state oil company. In the second type of sale, the state oil company satisfies its domestic needs and then sells whatever oil is left to other private oil companies, usually under one-year contracts. In the third, the state oil company reserves a portion of its share for sale on the spot market at whatever price it will bring or assigns some of it to favored individuals, who could be members of a royal family or ruling clique or others with close ties to the government. It is with these "others" that the broker deals.

Charter Crude wanted the additional Libyan crude badly enough to enter into a written agreement with Billy Carter that promised him a broker's commission dependent on market conditions at the time the company got its oil. If oil was plentiful, Billy would get 5 cents a barrel on 100,000 barrels. In a tight market, Billy's commission would be 55 cents a barrel—or roughly $20 million.

In late August, Billy flew to Tripoli where, as an unregistered foreign agent, he looked into various matters and also attended the celebration of the tenth anniversary of the Libyan revolution. Unfortunately for Billy, he returned to Buena Vista without any commitment from the Libyan National Oil Company to provide more oil to Charter. In fact, as part of a general cutback in its production, Libya reduced Charter's crude supplies on April 1, 1980, from 100,000 to 60,000 barrels a day.

The president's brother never did obtain for Charter the Libyan oil he had promised and thus did not get a dime of the projected $20 million in broker's commissions. The notoriety that attached to him after the news broke in July 1980 of his relationship with Libya would undoubtedly have disqualified him as a petroleum broker in any case, for the role requires secrecy, a low personal profile, and—above all—the ability to deliver the goods.

It is not to the brokers, the smaller independents like Charter, or the heirs of the old oil barons that one should look to appraise charges of profiteering at the public's expense. One should concentrate upon the majors and their chief executives, for they still dominate the American and international industry—or, as their critics would have it, they control and manipulate it.

It is important to understand that the men who now direct the majors are not colorful pioneers or bold entrepreneurs. They do not, like John D. Rockefeller, create huge corporations or, like Dad Joiner, discover mammoth oil fields on their own. The chairmen of the boards and chief executive officers of the majors are managers of enterprises created by their predecessors. They keep a low profile, except for an occasional speech or interview, and are not majority stockholders in the corporations they direct. The chairmen usually started out as engineers—not the best training for dealing with politicians and their bothersome regulations, the media and its bothersome questions about profits, or even Arab sheiks and their bothersome ways—and worked their way up through the ranks. Along the way, their salaries were relatively modest. They are not autocrats who own the companies they administer, but high-level employees who must account to stockholders, boards of directors, and government regulators. If the bottom line is deemed unsatisfactory, they can be ousted. At age sixty-five, they retire and are replaced by another graduate of the up-through-the-ranks school.

Typical of the current chairmen is Clifton C. Garvin, Jr., of Exxon. After graduating from Virginia Polytechnic Institute with a master's degree in chemical engineering, he joined Exxon in 1947 at its Baton Rouge refinery. He moved steadily up through the ranks, becoming president of Exxon's worldwide chemical business, president of the company in 1972, and chairman and chief executive officer three years later at the age of fifty-three.

A quiet, deliberate man, Garvin was elected chairman of the Business Roundtable in 1980 and as such became a spokesman for corporate America. He even appeared on a daytime interview program, the *Phil Donahue Show,* during the gasoline shortage of 1979. He has said of this novel experience: "I wasn't prepared for 5,500 women to literally boo at the answer I gave on why there was a gasoline shortage." After his TV appearance, Garvin had to switch to an unlisted home telephone number because he and his wife were getting so many abusive calls.

Is there any justification for this kind of hostility? The majors are undeniably earning enormous profits from the energy crisis and the soaring prices for petroleum products, but are executives like Garvin getting rich?

The year 1979 was a good one for the personal finances of the chairmen of the five American sisters. Corporate proxy statements reveal that Garvin was paid $830,000 in salary and bonuses, up 8.8 percent from 1978. In addition, he received $147,993 through employee profit sharing and various stock rights. Howard Kauffman, president of Exxon, received $525,000 in salary and bonuses.

The two top executives of the biggest oil company did not lead the field, however. That distinction went to Rawleigh Warner, Jr., chairman, and William P. Tavoulareas, president, of Mobil. Warner received $1,187,055 in salary and bonuses, up 27 percent from 1978, and Tavoulareas $1,005,482, up 25.5 percent. That was not their whole take. The Mobil executives also received "indirect" compensation by exercising stock options awarded in previous years. Warner realized $2.4 million in this manner and Tavoulareas $967,666.

At Texaco, Chairman Maurice Granville was paid $757,400 in salary and bonuses, up 14.9 percent, and President John McKinley got $522,600, up 28.3 percent. At Standard Oil of California, Chairman H. J. Haynes, was paid $529,583, up 8.8 percent, and President John Grey received $376,000, up 9.9 percent. Gulf had a curious situation: its chairman, Jerry McAfee, made $403,333 in salary and bonuses, up 16.1 percent, but less than its president, James E. Lee, who got $527,083, up 19.8 percent. However, McAfee received $164,169 in other compensation, which made his total compensation larger.

The chairmen of somewhat smaller oil companies also did not fare badly. For example, Fred Hartley of Union Oil of California received total remuneration of $843,054.

All these increases were largely tied to company profit gains. But while the figures indicate that oil company executives are doing quite nicely, when seen in perspective they do not sustain the charge of profiteering. It may be cold comfort to those paying rising heating oil bills to be reminded that Johnny Carson gets $5 million a year for his *Tonight* show appearances; Marvin Webster is paid $3 million spread over five years to play basketball for the New York Knicks; and outfielder Dave Winfield of the San Diego Padres signed a contract as a free agent with the New York Yankees in December 1980 that will reportedly pay him as much as $20 million over the following ten years. Big Oil's top executives are not in the same class with H.L. Hunt, the king of Saudi Arabia, or the late Shah of Iran. They are handsomely rewarded, but not extravagantly compared to the chief executives of other major industries with comparable responsibilities, sales, assets, and numbers of employees. For example, *Business Week*'s annual survey of executive compensation, which takes into consideration long-term income (LTI) benefits—mainly stock options and stock appreciation rights (SARs)—found that Mobil's Warner was only the second highest paid U.S. executive in total compensation in 1979. The first spot went to Frank Rosenfelt, president and chief executive officer of Metro-Goldwyn-Mayer, whose total compensation was more than $5 million, of which $4.9 million came from the exercise of stock options. Mobil's Tavoulareas came in seventh among all executives, and Exxon's Garvin only twenty-fifth.

Now we come to the heart of the controversy: Are the oil companies themselves making exorbitant profits at the public's expense? Here we must enter a world of dollar figures so huge as to be daunting, bearing in mind that the oil companies have perfected "creative accounting" techniques that make their financial statements hard to penetrate.

In *Fortune*'s list of the five hundred largest U.S. industrial corporations in 1979 as ranked by sales, six of the ten largest and nine of the fifteen largest were oil companies (AT&T is a public utility and therefore is not considered an industrial corporation). Exxon led the list, with gross revenues of $84.35 billion, more than any other business had ever reported for a single year, and a record net income (profits) of $4.3 billion. General Motors was second, followed by Mobil (with sales of more than $44 billion); Ford; Texaco (sales of more than $38 billion); Standard Oil of California (sales of more than $29 billion); Gulf (sales of more than $23 billion); IBM; General Electric; Standard Oil of Indiana (sales of more than $18 billion); ITT; Atlantic Richfield (sales of more than $16 billion); Shell Oil (sales of more than $14 billion); U.S. Steel; and Conoco (in fifteenth place with sales of more than $12 billion). Even the smaller independents prospered. Amerada Hess, for instance, with sales of more than $6 billion, placed forty-first on the Fortune 500 list.

These sales and net income figures, coming in a year when consumers were desperately coping with both serious shortages and higher prices, aroused a great deal of anger. Representative Norman Mineta, Democrat of California, summed up the reaction of many of his colleagues when he said: "The public really feels it's getting ripped off." An unusual protest was registered by Elizabeth Hird Rauch of Killingworth, Connecticut, a Texaco shareholder. Her 1979 second-quarter dividend was $342.90, $25.40 more than her dividend for the first quarter. She returned the $25.40 to Texaco, together with a letter to its chairman, Maurice Granville: "I do not wish to profit from the current shortage and your policy of taking advantage of it to raise prices of gasoline and other Texaco products." Harvey Rosenfeld, a lobbyist for Ralph Nader's Congress Watch, observed: "After these awesome profits the question is not whether we should have a windfall profits tax but whether we should decontrol domestic prices at all."

At a news conference called to discuss Exxon's 1979 performance, Clifton Garvin remarked: "I guess AT&T will be the only company that will have higher profits, in dollars, this year than us." Though he conceded that Exxon seemed to "prosper while the public suffered," Garvin staunchly maintained that the company's profits were not excessive when measured by the yardstick of return on shareholder's equity.

There is a difference between *profits* expressed in terms of absolute dollars of net income and *profitability*. The latter is the relationship between the profits that a company clears and the investment needed to generate them. Equity is a company's assets minus its liabilities. Return on equity is the profit expressed as a percentage of the equity, or how much is being earned by the money that's been invested in the business.

Exxon's rate of return on equity rose from 14 percent in 1978 to 20.1 percent in 1979; Mobil's was 20.8 percent. But return on equity for the broadcasting and motion-picture production and distribution industry was higher at 22.2 percent, while the return on equity for all industrial companies, according to government figures, averaged 16.7 percent. Also, the rate of return for the oil companies was slightly lower than it had been for the five years prior to 1979. So a case could be made that the profits of Big Oil are not excessive in relation to what companies in other industries earn on investment. Of course, this is precisely the case oil company executives are so fond of making.

This argument, however, is economic sophistry, for it leaves at issue a question that can't be answered by statistics. Can an Exxon, with sales of $84.35 billion and net income of $4.3 billion, really be compared to another international company such as Pepsico, with sales of a bit more than $5 billion and net income of about $265 million, even though Pepsico's rate of return on stockholder's equity was 25.6 percent and Exxon's was "only" 20.1 percent? Does Big Oil, which dominates this nation's energy resources and commands such enormous amounts of money through sales and net income and still enjoys unique tax breaks, really need to make substantially greater percentage gains than other industries?

Even granting that the profits of the oil companies, when expressed as return on stockholder's equity, are pretty much in line with the national average for large manufacturing companies, two questions still arise. Why did their total and net revenues increase so much in 1979 compared to 1978, and why did consumers have to pay so much more for gasoline, heating oil, and other petroleum products while this was happening?

One answer to the first question is that profits in 1978 were somewhat flat. Had they not been, the increases registered in 1979 would not have appeared so steep. Another is "inventory profits." These are perfectly legal and arise when companies revalue the stocks they have on hand to reflect higher OPEC prices. They pay taxes on the profit, but then they also have to replace the stocks at the higher prices. "Inventory profits" are a temporary phenomenon.

The multinationals also maintain that a large part of their

profits (80 percent of their growth, according to the American Petroleum Institute) came from overseas, especially Europe, where fewer price controls and a very cold winter permitted the companies to boost prices quickly to meet higher demand. Gulf even bought full-page newspaper ads to proclaim: "Almost all of this year's first-half increase in net income occurred outside the United States," the implication being that those big profits were not extracted from the American consumer. But like almost everything else about the complex, secretive oil business, the issue is not quite so simple.

The multinationals do normally attempt to maximize their foreign income to take advantage of foreign tax breaks. This is especially true in times of large price increases. However, there is nothing to prevent a company from selling expensive crude oil it produces abroad in the United States, where it is refined and marketed. The transaction is chalked up to foreign operations, keeping U.S. profit margins, and therefore taxes, low, though the product has been bought by American consumers. It is difficult to determine how often these intracompany sales take place because the annual reports of the majors usually aren't broken down in enough detail to make the determination.

In any case, since the oil business is cyclical, foreign and inventory profits are not likely to continue at such a high level in the 1980s. Of more interest to consumers than these arcane financial matters is the simple question: How did prices get so high?

In 1977 and early 1978, there was a worldwide "glut" of oil—an excess of production capacity. The producing nations had both the desire and the capability to produce more oil than the world wanted, so the price remained fairly stable. By the middle of 1978, however, demand began to overtake supply, and in the fall of that year came the revolution in Iran, which had been supplying about 10 percent of the noncommunist world's oil needs and 5 percent of America's. Iran's exports stopped completely in the first two months of 1979, then resumed at a reduced rate. The revolution in Iran did not in itself cause gasoline shortages and price rises in the United States, it only triggered the cycle. Other OPEC members, mainly Saudi Arabia, increased their production to bridge the gap created by the loss of Iranian oil, but never closed it completely. The result was a tight market vulnerable to price rises.

OPEC seized the opportunity at its meeting in Geneva in June 1979 to raise prices by 50 percent over the previous December level, when Iran began cutting production. There were two new prices: Saudi Arabia, Qatar, and the United Arab Emirates agreed to sell oil at $18 a barrel, while the other members were allowed to go as high as $23.50.

The major companies had no choice but to pass these increases along, since neither they nor the U.S. government could influence OPEC prices. After the addition of tanker costs; profit margins for the refiner and wholesaler; heavy federal, state, and local taxes; and shipping costs by barge and truck to service stations that, in a tight market, raise their profit margins to the maximum legal limit, the price of gasoline went over $1 a gallon.

The price of home heating oil also increased dramatically; in the Washington, D.C., area it rose from 49 cents a gallon in August 1978 to 83 cents a gallon a year later and continued to rise. Heating oil is refined from the same crude oil as gasoline; 20 gallons of gasoline and 10 gallons of heating oil, along with about eight other products, can be made from one barrel of crude. Unlike gasoline, heating oil was not under price controls. Its rise in price reflected the truism that the consumer winds up paying more whenever a shortage of crude exists because all oil prices go up to the highest price the market will bear.

No evidence has been found to support the widespread suspicion that the oil companies hoarded supplies of gasoline and heating oil in anticipation of higher prices in order to make a killing at the consumer's expense. Still, they benefited from the tight market and higher prices created by OPEC because profit margins tend to widen whenever supplies tighten in any business, although U.S. price controls somewhat checked this trend in the oil business.

This neat explanation is not the end of the story. In the

complicated mosaic of oil economics, the interests of the Seven Sisters, and some of their smaller competitors, more often coincide with those of OPEC than with those of the American consumer. This is not to say that the companies actively conspire with OPEC, but whenever the cartel jacks up the world price, the value of the companies' holdings in non-OPEC areas like Canada and the British and Norwegian North Sea go up. So it would not be in the companies' financial interests to pressure OPEC to hold down prices, even if they could still exert such pressure.

Another intriguing piece of the mosaic has to do with Saudi Arabia, which produces one-third of OPEC's oil and 20 percent of America's imports. Almost all Saudi oil is produced by the four Aramco partners, Exxon, Mobil, Texaco, and Standard of California. In 1979, the moderate Saudi government raised the price of its main grade of oil, Saudi light, by 42 percent over the price at the end of 1978, but still it cost from $2 to $4 a barrel less than comparable crude oil from the more avaricious OPEC members. However, the Aramco partners generally sold petroleum products refined from Saudi oil at about the same price as their competitors charged for products made from higher-priced oil. The result had to be a bonanza, the exact size of which is buried in the financial statements of the Aramco partners. Nevertheless, a rough guess can be made. Assume that the partners produce 6.5 million barrels a day at an average price advantage of $3 a barrel. That would come to potential additional profits of $19 million a day (before taxes), or about $7 billion a year (before taxes).

Taxes are still another interesting piece of the mosaic of oil economics. Little understood provisions of the U.S. tax code save the oil majors an estimated $1.2 billion a year. In theory, the foreign tax credit, which applies to all American companies operating abroad, seems reasonable enough. It is a dollar-for-dollar credit that the U.S. government allows for income taxes paid to foreign governments and was designed to ensure that no company would be taxed twice on its foreign income. In practice, though, the credit is particularly beneficial for the oil companies. Understanding this requires entering a world of financial legerdermain where the distinction between "royalties" and "income taxes" becomes elusive.

Oil-producing countries naturally expect to be paid for oil pumped within their borders. Such payments are generally called royalties, as are the payments made to private American citizens who own oil reserves. In the late 1940s, Saudi Arabia was getting a mere 12 percent royalty from Aramco. Then Venezuela negotiated a fifty-fifty split for itself, which made King Ibn Saud unhappy. To mollify its principal ally in an insecure part of the world at the height of the Cold War, the U.S. government agreed to an ingenious scheme that would enrich Saudi Arabia but not harm Aramco's profits. The Saudis were allowed to take a bigger piece of the pie by imposing a corporate "income tax" that would be certified as eligible for the foreign tax credit. The payments would reduce the companies' U.S. taxes dollar for dollar. As royalties, these payments would have been considered a cost of doing business (which they really were), and as such they would only be deductible like any other business expense and thus worth less than half as much on the companies' tax returns.

The companies were content with this arrangement and so were the Saudis, whose new "income tax" took effect on December 30, 1950. Within a year, Aramco payments to the Saudis leaped to $110 million from $66 million, while its payments to the U.S. Treasury dropped from $50 million to $6 million.

It was only after the Arab oil embargo of 1973–1974 that this arrangement excited the curiosity of some U.S. legislators. In 1974, the following dialogue took place in testimony before the Senate Subcommittee on Multinational Corporations between Senator Frank Church and George McGhee. A former vice president of Mobil, McGhee had been assistant secretary of state for Near Eastern, South Asian, and African affairs in 1950.

SENATOR CHURCH: If an oil company were doing business within the United States and had to pay a royalty to the owner of the land on which the wells were located and had to pay taxes to the state in which the lands were situated, both the royalty and the taxes paid to the state government would be treated, if I understand the law correctly, as regular business expenses and deducted as regular business expenses in determining the profit on which the company would have to pay income tax to the Federal Government.

But upon the recommendation of the National Security Council, the Treasury made the decision to permit Aramco to treat royalties paid to Saudi Arabia as though they were taxes paid to the Arabian Government. The effect was dramatically different from the U.S. example because instead of deducting those royalties as regular business expenses, in determining the net profit, Aramco was permitted to credit those royalties directly against any tax otherwise due the Government of the United States.

So that, first of all, the impact on the national treasury was direct and dramatic and resulted, I am told, in a loss of over $50 million in tax revenues from operations of Aramco in Arabia.

Isn't that true?

AMBASSADOR MCGHEE: Yes. I might elaborate a little. I find in my records that I pointed out at the time, that the U.S. Treasury would in a sense be financing this change, and I didn't make the decision to give the tax credit. . . .

SENATOR CHURCH: However, even then the effect of the decision was to transfer $50 million out of the U.S. Treasury and into the Arabian treasury. That was the way it was decided to give Arabia more money and to do it by the tax route. Isn't that correct?

AMBASSADOR MCGHEE: Yes, that is one way of looking at it. . . .

SENATOR CHURCH: In this case the reason that I think it is so intriguing is because the companies had been paying a royalty that received one tax. Then a decision was made in our Government to treat that royalty differently so it would have the status of a tax credit with the effect that $50 million in the next year was transferred out of our treasury into the treasury of Saudi Arabia.

AMBASSADOR MCGHEE: That is right.

Senator Charles Percy found all this a little hard to believe:

> SENATOR PERCY: You would think there would be screams from the Treasury that is always looking for revenue. . . . How much protest was there from the Treasury and where did it come from and who resolved the difference and the difficulty then?
> AMBASSADOR MCGHEE: I frankly don't know, Senator. . . . Having made my recommendation about the political and strategic aspects of the question I frankly left it to Treasury to make this decision. But the impression I had then was that no one objected to it. Everyone thought it was reasonable under the circumstances.

The circumstances were that the foreign tax credit was designed to placate Saudi Arabia and hold down the oil companies' tax liability. Let us say an oil company pays $1 to the Nigerian government as royalties. That reduces the company's taxable profits in the United States by $1, saving about 46 cents in corporate taxes. However, to avoid double taxation, income tax payments to foreign governments are directly credited against American tax liability. So let us say that the same $1 is given to the Nigerians as income tax instead of royalty. The company then reduces its American taxes by $1 instead of 46 cents.

The Nigerians, Saudis, or any of the other oil producers don't care whether their payments are called royalties or income taxes, so long as they get paid. Neither do the companies, so long as they maintain the profits made possible by such preferential taxation. The American taxpayer, who has to make up the difference, is the only loser in this game of semantics. In 1978, Aramco earned profits of more than $580 million, on which it paid no U.S. income taxes whatever. Thus one gains an insight into why oil company accountants toil long and hard to find ways to convert royalties into income taxes.

Multinational oil companies do not disclose what they pay in "income taxes" to the oil-producing countries. But the experience of Exxon is instructive. In 1978, the world's largest oil company had pretax earnings of $2.40 billion in the U.S. and $6.05 billion overseas. Exxon reported paying $29 million in federal taxes on its foreign income, and $4.63 billion in foreign income taxes—an effective rate of more than 76 percent. On its domestic income, Exxon paid taxes of $1.03 billion—a substantially lower effective rate of 43 percent. The result was that Exxon's domestic *net* income of $1.37 billion almost equaled its net income of $1.39 billion from overseas, even though its *gross* income from overseas was two and a half times its gross domestic income.

In a study for the Ford Foundation, Gerald Brannon of Georgetown University concluded: "There is every reason to assert that the bulk of oil company payments to host countries are in fact royalties." Brannon terms the federal tax credit for such payments "highly questionable." Jack Blum, a Washington attorney who served for eleven years as a staff member of the Senate Antitrust and Monopoly Subcommittee and the Foreign Relations Committee, observed: "We have reached the point with the oil companies where the foreign tax credit is being abused on a scale that no one had imagined. The whole scheme is now simply subsidizing foreign imports."

In November 1980, the Treasury Department finally proposed new Internal Revenue Service regulations changing the foreign tax credit. For one thing, the regulations would disallow a tax credit for taxes in nations that impose a charge on net gain of an oil company using a posted or artificial price. One would have to be a certified public accountant to understand the ramifications of the new regulations, but the furious reaction to them on the part of the oil companies was predictable (the five majors alone claimed $18 billion in foreign tax credits in 1977).

In an op-ed page ad Mobil said: "The new rules, denying to U.S. oil companies the foreign tax credits that all other U.S. companies get and all major industrial countries give, would saddle the incoming Administration with a burden that could render our petroleum industry uncompetitive with foreign companies in obtaining crude oil supplies abroad." The industry's position was that it must have such tax breaks, in addition to high profits, to attract investors and generate the money to finance domestic exploration. In fact, the majors complained in a barrage of ads aimed at the suspicious consumer in 1980 that their operations are widely misunderstood and that, far from enjoying bloated profits, they aren't making *enough* money to invest in new production in order to reduce the nation's oil imports.

From Shell:

> Surprising but true. That 4.1 cents represents the average per gallon profit on all of Shell's petroleum products and natural gas sales. The total dollar profits reported by oil companies seem high mainly because oil companies sell an awful lot of gallons. . . . Present and future profits allow us to put money back into our business and attract investors so we can make the huge investments necessary to help solve America's energy problems.

From Chevron (Standard Oil of California):

> Actually, Chevron's 1979 profit on each dollar of U.S. petroleum sales was 5.1¢. (This compares to a 9-month average of 5.6¢ for all other major U.S. industries.) Part of Chevron's profit, of course, went back to our shareholders. The remaining profit and other cash from operations provided the funds for Chevron's expenditures in such areas as exploration and development of oil and gas fields, refineries, and transportation facilities.

Chevron concluded this ad, and others like it, with the phrase: "Thank you for listening."

Exxon did not thank us for listening but earnestly explained:

> Exxon's profits are large but so are its investments in energy. When Exxon's profits are put on a comparable basis with the profits of other U.S. manufacturing companies, we do only a little better than average. . . . Oil companies need to be at least as profitable as other companies if they are to continue to be able to attract the funds necessary for the high risk, big investments vital to the delivery of energy in the future.

In another ad, Exxon portrayed itself as a company owned by worthy institutions and average Americans instead of a heartless conglomerate:

> Millions of Americans have either a direct or indirect share in Exxon profits in the form of dividends. Beyond the 625,000 individuals who have a direct ownership in Exxon, there are 60,000 institutions such as pension funds, trust funds, colleges, foundations and insurance companies in which millions of people have a stake. In 1979, Exxon's total dividends will give our shareholders a return of about 7% on the current price of Exxon stock.

What about these shareholders? Surely these Americans, at least, benefited from higher petroleum prices. Exxon did hand out $1.7 billion to its shareholders in 1979, up from $1.5 billion in 1978, while Standard Oil of Ohio doled out $443 million, up from $410 million. The American Petroleum Institute, in its study *Who Owns Big Oil?*, claimed that the leading six oil companies had fourteen million shareholders. Bob Hope, the TV spokesman for Texaco, assured his viewers that lots of ordinary Americans own Big Oil and benefit from its profits.

But like almost everything else in the oil game, the matter of ownership is more complicated than the companies allow. The cheering image of millions of widows, orphans, and retired Americans supported by oil stock dividends is shattered when one realizes that of those fourteen million owners, only 2.3 million are direct individual owners who can sell their stock whenever they choose. The rest are indirect owners through such institutions as insurance companies, mutual funds, and private or public pension plans. Principal ownership, and voting control, of Big Oil stock is concentrated in the hands of a few wealthy individuals and powerful institutions; the rest is spread thinly among all those folks Bob Hope talks about. Upon Texaco's board of directors sits Sir Arthur Patrick H. Forbes, the Earl of Granard, who lives in Morges, Switzerland, and owned 113,018 shares of the company's stock as of January 1, 1980. Those shares were worth roughly $4,294,684 in July 1980. Another member of the board is George Parker, Jr., of San Antonio, Texas, who owned 121,036 shares worth about $4,599,368.

Presumably representing the "typical" owner are two other board members: Dr. Lorene Rogers, president emeritus of the University of Texas at Austin, and Dr. William McGill, former president of Columbia University, each of whom owned 200 shares worth a mere $7,600. One wonders how much weight the opinions of Drs. Rogers and McGill carry at Texaco's board of directors meetings in contrast to those of Lord Granard and Mr. Parker.

According to a 1980 study by the New York City–based nonprofit research organization Corporate Data Exchange Inc., major banks and other large financial institutions dominate stock ownership of the nation's largest energy companies. Fewer than fifty financial institutions control at least 15 percent of the stock in at least the thirty-eight leading energy companies. Among these financial institutions are the Chase Manhattan Bank, Bankers Trust, Citibank, and Morgan Guaranty; the last is the second largest holder (2.08 percent) in Mobil. Other large stockholders are pension funds, such as the Teachers Insurance and Annuity Association; insurance companies, such as the Prudential Insurance Company of America; and investment companies, such as Capital Growth Inc. In widely held public companies like the oil majors, shareholdings that represent only a small percentage of the total stock outstanding are often enough to guarantee significant control.

The relationship between the large oil companies and the major banks finally aroused the curiosity of the Federal Trade Commission, which began an investigation in the summer of 1980 with the purpose of determining what effects this relationship has on competition in the petroleum industry. Privately held companies, like those of the Hunts, are not subject to such bothersome scrutiny.

Which brings us to the crucial question about the oil company money game: What exactly is the industry doing with the enormous amounts of money it is taking in? Now that all price controls on U.S. oil production have been removed as of late January 1981, there will be a further sharp increase in oil company profits, the windfall profits tax notwithstanding. Will those accelerated profits lead to the increased investment in domestic exploration that the federal government and the industry both maintain is necessary, and if so, what will be the likely result?

The industry is united in its claim, as one Mobil ad put it, that "Big as our profits are we consistently spend more than we earn to find and develop new sources of energy." Exxon planned to spend $6.6 billion on energy projects in 1980, including $3 billion on oil and gas projects in the United States, and $1.2 billion in other Western Hemisphere countries. Exxon also planned to spend $1.4 billion on other activities, including coal, chemicals, and uranium. Gulf planned to spend a record $3 billion for capital and exploration projects for 1980, a 25 percent increase over the $2.4 billion it spent in 1979. Atlantic Richfield reinvested $1.4 billion in its petroleum activities in 1979 and also spent $100 million on expanding its coal-mining and solar-energy subsidiaries.

But these lavish promises have not placated the industry's critics. James Flug, a leading consumer lobbyist who directs the Washington-based Energy Action group that monitors the industry, said: "Beyond a certain level higher prices just bring higher profits, not more oil." Edwin Rothschild, also of Energy Action, commented: "What is really important is cash flow. The oil companies' cash flow is about 50 percent higher than profits, and they are not spending it all on the search for oil and gas."

And Senator Howard Metzenbaum of Ohio, a persistent industry critic, wrote that the combined cash flow of the major oil companies in 1978

> was an astounding $26 billion. In addition, decontrol is expected to bring the industry an extra $49 billion to $80 billion by 1985 in after-tax revenues on top of its exceedingly high profits. . . . The fact is that big oil is not doing all it can to develop oil and gas resources. Exploration expenditures by the top eight oil companies in 1977 were about the same as those by the very smallest oil companies. And the industry spent only 8.6 percent of its profits on research and development, a small fraction of what other industries spent. . . . The real-world fact is that big oil companies rely significantly less on the capital markets than they do on their own huge internal cash flow. Even for the external capital these 16 companies do need over the next 10 years, they don't need to diversify. The big oil

companies will remain enormously profitable investments during the next decade. . . .

Whether one sides with Senator Metzenbaum and Energy Action or with the oil companies, one fact is beyond dispute: Spurred by decontrol and higher prices, exploration and drilling are going on at record rates. Perhaps, then, the oil companies' profits are acceptable if they result in increased U.S. production. But will they?

Some new oil is being found, and Alaska has great potential, but despite the huge sums being invested in exploration and drilling, crude oil production in the lower forty-eight states continues its long decline. "Unfortunately, as prices get higher, production forecasts get lower. It is peculiar economics," commented Lawrence Goldstein, senior economist for the Petroleum Industry Research Foundation in July 1980. Paul Hoenmans, executive vice president of Mobil's exploration and production division admitted that "It doesn't appear there are great volumes of oil yet to be found in the U.S.," and Robert Baldwin, president of Gulf's refining and marketing unit, observed: "If crude oil went to $100 a barrel, we couldn't arrest the decline rate."

So while the oil companies are being castigated for making huge profits and counter defensively that they need those profits to look for more oil, it appears there is not much oil left to be found. It is an irony that provides cold comfort.

It would take something of a miracle to boost U.S. production enough to relieve the nation's dangerous dependence on imports. Of course, miracles like Spindletop have happened in the oil game before, and estimates of low reserves have proved wrong in the past. Optimistic projections of stable prices based on market sense have also proved wrong. Milton Friedman, distinguished University of Chicago economist, Nobel laureate, and advocate of the free market, predicted in 1975 that OPEC would never raise the price of its oil to $10 a barrel because that would ruin the cartel. By June of 1980, OPEC's prices averaged $32 a barrel, and Algeria, Nigeria, and Libya set prices of $37 a barrel for their high-quality crude as of July 1.

The controversy over prices, profits, and new domestic supplies will only be resolved by time. For now, though, one thing is absolutely certain: The American people need the expertise of the majors, the big independents, and the wildcatters to find and supply whatever oil is left as we make the transition to serious conservation and other forms of energy.

It would be comforting to believe that the first priority of the men who run the oil business is to ensure adequate supplies of reasonably priced petroleum products during the transition period, but it is more realistic to assume that their first priority is to make money by providing consumers with certain goods that they want to buy. If they can also satisfy our energy needs without depleting our family budgets, they would consider that a goal worthy of reaching.

Chapter Five

Who's Running the Show?

Until the energy crisis threw an unwelcome spotlight on their industry, the men who run the oil business were as faceless as members of the Soviet Politburo or the Libyan Revolutionary Command Council. Even today, the men (and they are all men) who direct the world's biggest industry are hardly household names or recognizable faces to Americans who depend on them for three-quarters of their energy.

These men are not all alike, except in their shared devotion to the bottom line and aversion to government regulation. The careers of some of them, selected at random, illustrate the often startling diversity that can be found in the exccutive suites of Big Oil.

The Quiet Oil Man

You will not find Leon Hess sipping a bourbon and branch with the good ol' boys at the Houston Petroleum Club, ingratiating himself with Wall Street analysts, or granting informative interviews to the business press. He is a loner, a fiercely independent entrepreneur who started from scratch, took on the giants, and became one himself. His ascent has been marked by keen business judgment, controversy, a generous dose of *chutz-*

pah, luck, an aversion to the limelight, and skillful manipulation of some curious government policies.

Leon Hess did not get his start in the oil business as a Louisiana wildcatter, Oklahoma roustabout, or geology student at Texas A & M. He began in 1932, at age eighteen, by driving a delivery truck for his immigrant father's small fuel oil business in Asbury Park, New Jersey, his birthplace. He began to expand the family business just before World War II with a characteristically unusual tactic. Hess bought nine old 20,000-barrel tankers and anchored them next to an abandoned brickyard in Perth Amboy, New Jersey, as a storage terminal.

After the war, which he spent as a petroleum supply officer with the 3rd Army in Europe, Hess continued to expand the family business by developing a network of terminals and tank trucks on the northeast coast. As a regional fuel oil distributor, he was hardly a competitive threat to the Seven Sisters. But he first began to buck them by buying his own tankers and fuel oil on the open market, instead of following the standard custom of purchasing product delivered in the majors' vessels under long-term contracts.

In 1958, Hess built his first refinery near Perth Amboy (now closed). The next year, he got into gasoline marketing on the northeast coast with innovative departures from the traditional practices of the majors. Hess helped to pioneer the high-volume discount service station that sold only gasoline, offered no maintenance services, and accepted only cash. A typical Hess station pumped three times the average volume of the stations of his bigger competitors—some 180,000 gallons a month—and undersold them by a few cents a gallon. The high-volume no-service concept is now being increasingly adopted by the majors.

In 1963, the family-owned business went public, although Hess retained 21 percent of the stock and continued to run it as a one-man show. Three years later, he took his biggest gamble by building the world's largest refinery in the Caribbean, at St. Croix, Virgin Islands. The choice of St. Croix might have seemed odd on the surface, but Hess was capitalizing on the territorial government's guarantee of exemptions from import fees and property taxes as well as a rebate on income taxes that gave his still modest company an effective tax rate of 12 percent. But these concessions were not the only reason why the choice of St. Croix turned out to be a master stroke.

The Jones Act of 1920 stipulates that U.S.-flag vessels must be used in trade between U.S. ports. The American Virgin Islands are U.S. territory, but they are exempt from the Jones Act. Thus Hess, though really a "domestic" refiner, was legally free to ship refined products from St. Croix to his chief market, the East Coast, in foreign-flag tankers. The voyage from the Caribbean in foreign-flag tankers cost about $1.35 a barrel. A trip of about the same length from Gulf Coast refineries to the East Coast in U.S.-flag vessels cost the majors $3.50 a barrel.

That advantage enabled Hess to become the biggest supplier of residual fuel oil on the East Coast. He captured about 20 percent of the market at the expense of majors like Exxon, Texaco, and Shell and smaller independents who had not appreciated the nuances of the Jones Act and preferential tax breaks. Hess's strategy concentrated on the often-ignored market for residual fuel oil, which is a heavier oil used by gas and electric utilities, industrial plants, and large apartment and commercial buildings. Almost 60 percent of the St. Croix refinery's production is devoted to residual fuel, with the remainder split between the lighter heating oil and other distillates and gasoline. Most refineries produce about 45 percent gasoline and 15 percent residual fuel.

Hess, though now engaged in refining, transportation, and marketing, was still not a truly integrated oil company because it had no large producing operation; its production amounted to only 3,500 barrels of crude a day in Mississippi. The majors, of course, had all begun on the producing side. Hess reversed the usual practice by integrating backward into production. In 1969, after three years of often acrimonious debate with its chairman and chief executive officer, who died that year, Hess managed to acquire the Amerada Petroleum Corporation.

Amerada was a leading independent crude oil producer, pumping 220,000 barrels a day, mainly in North Dakota and Texas. Moreover, Amerada had a half interest in 777 million barrels of reserves in Libya and owned proved reserves of 582 million barrels in Canada and the United States. In partnership with Standard Oil of Indiana, Amerada also owned tracts in the North Sea. Oil was discovered on them in 1971 and had reached an output of about 30,000 barrels a day by 1979. Amerada Hess now has a major position in three producing fields in the North Sea and is drilling in four others.

The timing of the merger couldn't have been more prescient. In 1969, supplies of crude oil were plentiful and producers complained of weak prices. Then, with the Arab embargo of 1973–1974, followed by stunning increases in world oil prices and gradual decontrol of domestic oil prices, Amerada Hess's reserves began to resemble black gold.

In 1979, Amerada Hess posted a net income of $507,116,000 on sales of $6,769,941,000 and ranked as the fifteenth largest American oil company. But Leon Hess did not become part of the petroleum industry establishment. A tall, lean, white-haired man who dresses well, he lives in an apartment on Manhattan's Park Avenue and works long hours at his corporate headquarters in Rockefeller Center. His wife serves on the board of directors of the Lincoln Center for the Performing Arts. He has two daughters and a son, John, who is in his late twenties and is a director of Amerada Hess with the title of "coordinator of planning and control."

Hess himself shuns publicity and declines to talk to the media or financial analysts. His company not only doesn't have an investor relations department, it doesn't even have a public relations department at a time when Exxon, for one, is spending 78 percent of its $18 million advertising budget on promoting the idea that its interests and those of the American people are coincidental.

Hess is rarely photographed, although he was in 1972 in the locker room of the New York Jets with Coach Weeb Ewbank, Joe Namath, and Henry Kissinger. Hess is chairman and chief executive officer and 75 percent owner of the professional football team; it is one of his few outside interests.

Because of his lone wolf style, Hess has a number of detractors in the industry who suspect that he has somehow used political clout to gain competitive advantages from U.S. government policies. He is a Democrat in a business whose executives are usually Republicans, and his father-in-law, David Wilentz, was once a Democratic national committeeman from New Jersey. Hess himself was close to Stewart Udall, who was secretary of the interior when the department awarded Hess a favorable oil import allocation from the Virgin Islands. Hess also contributed $225,000 to Senator Henry Jackson's unsuccessful 1972 bid for the Democratic presidential nomination. Jackson happened to be chairman of the influential Senate Energy Committee.

While Hess's critics are undoubtedly motivated more by envy than by any evidence of wrongdoing, his St. Croix refinery has enjoyed the best of both worlds in regard to the "entitlements" program. This program was established in 1974 to equalize the costs between what domestic refiners paid for price-controlled U.S. crude oil and the more expensive OPEC supplies. Hess's Virgin Islands plant was considered a domestic refinery, but had little access to U.S. crude oil, being mostly dependent on supplies from Iran, Libya, and Abu Dhabi. So under a plan financed by refiners with access to lower-cost mainland crude, Hess's St. Croix plant received $881.6 million in subsidies through July 1979.

While being treated as a "domestic" refiner in terms of the entitlements program, Hess was treated as a "foreign" refiner in terms of the Jones Act and thus could use the lower-cost foreign-flag tankers to undersell his competitors. Their outrage forced Congress to reduce Amerada Hess's subsidy in 1978; the entitlements program is scheduled to be phased out entirely by September 30, 1981.

The favorable Virgin Islands concessions Hess obtained are also due to expire in 1981, and the territorial government is planning to reduce them after that. From the time Amerada Hess was first lured to the islands through the summer of 1979, some $35 million in property taxes and import fees has been waived by the government, which also gave the company a hefty rebate in income taxes.

Not unnaturally, Hess strongly objects to any change, and he has bargaining leverage. He announced plans to shift up to 40 percent of the St. Croix refinery's operations to St. Lucia, a newly independent nation a few miles off the sea lanes to St. Croix where Hess has built a 100,000-barrel-a-day refinery as well as a large oil transshipment terminal. The St. Lucia refinery could easily be expanded. In return, the island's government has given Hess a fifty-year license providing exemptions from franchise, property, and income taxes.

Hess has thus confronted the Virgin Islands government with a dilemma common to Third World governments. It needs to raise revenue to bolster its faltering economy and close its huge budget gap. Short of raising its citizens' taxes, about its only means of increasing revenue is to collect property, income, and franchise taxes from Amerada Hess and Martin Marietta Aluminum, St. Croix's other major employer. But at the same time, it badly needs the jobs provided by the Hess refinery and the $18 million container port the company is building on Limetree Bay at St. Croix. If the government pushes too hard to collect the taxes, it might find the entire Amerada Hess operation located across the blue, sun-drenched waters in the more economically hospitable climate of St. Lucia.

Only Leon Hess knows for sure what he will eventually do, and as usual he is keeping his own counsel. His career proves that it is still possible—though certainly not easy—for an independent starting out from scratch to break into the unified ranks of Big Oil.

So, too, does the more flamboyant career of Dr. Armand Hammer, whom no one has ever accused of keeping a low profile.

The Oil Man Who Knew Lenin

If Leon Hess is the most reclusive of major American oil men, Dr. Armand Hammer is the most unlikely. He is also the oil man who, though not by intent, is most responsible for placing the American consumer at the mercy of OPEC.

In a rather colorless executive age, Hammer's personal style is a throwback to that of the legendary Texas wheeler-dealers, though he resembles them in no other way. A short, stocky man whose grandfatherly charm conceals a wily toughness, Hammer didn't even get into the oil business until he was fifty-eight years old and had retired from a number of successful careers including hustling pharmaceuticals, pencils, furs, jeweled Easter eggs, Old Masters, bourbon, and fertilizer.

Armand Hammer was born in New York City on May 21, 1898. His great-grandparents had emigrated there from Odessa, the warm-water trading port on the northern coast of the Black Sea. His father, Julius, practiced medicine in the Bronx and was a founder of the American Communist Party. Julius also spent two and a half years in Sing Sing for performing a fatal abortion, but was pardoned by Governor Al Smith and readmitted to medical practice.

In 1919, Armand Hammer got a B.S. degree from Columbia College and then graduated from medical school near the top of his class. But he never practiced medicine for any length of time, being more attracted to the wheeling and dealing of the business world. He was especially fascinated by the potential for doing capitalist business with the new Soviet government. His first profitable deal was to sell the Bolsheviks pharmaceuticals despite the postwar Allied blockade.

In 1921, Hammer went to Russia where, aided by his father's political contacts, he arranged a barter deal of Soviet precious stones, caviar, and furs for a million tons of American wheat. He met Lenin, who was eager to attract Western capital for Soviet development, and was allowed to form a private export-import firm that became the agent in the Soviet Union for almost forty foreign companies, including U.S. Rubber, Parker Pen, and Ford Motor. The Brown House, a twenty-four-room mansion with eight servants, was Hammer's base of operations in Moscow.

Hammer had no interest in communist ideology; he was in Russia to indulge the controlling passions of his life—making deals and making money. When the Soviets took back control of their foreign trade in 1925, they allowed Hammer to open a pencil factory with the help of German technology. He claimed that the pencils, sold as far away as China, made millions, split fifty-fifty between himself and the Soviet government. However, the good fortune of this resourceful capitalist did not outlast the rise of Stalin and his program of self-sufficiency. A. Hammer Pencil Company was bought out, the factory was renamed the Sacco and Vanzetti Works, and the entrepreneur left the Soviet Union in 1930. Hammer went on to conquer a number of fields before he got into oil in a big way.

His first move was to establish a bank in Paris to exchange Soviet promissory notes. A year later, at the urging of his younger brother Victor, who had been in Russia with him, he returned to New York City. Victor had brought back a great number of art objects and paintings from Russia, but they weren't of top quality and he couldn't unload them during the depths of the Depression. Armand had the idea of selling them through department stores like Marshall Field of Chicago, using bargain-basement promotion techniques.

The idea was a smashing success and the Hammer brothers went on to sell a substantial portion of William Randolph Hearst's huge art and antique collection and Franklin D. Roosevelt's family items. Armand Hammer later bought the Roosevelt home at Campobello and donated it to the American and Canadian governments as an international park. The Hammer Galleries in Manhattan are still prospering, although now they sell contemporary and Impressionist art rather than icons and the Czar's jeweled Easter eggs.

In the 1930s, Hammer branched out from art to beer barrels, which were in short supply with the end of Prohibition. He became the leading manufacturer of beer barrels, importing the wood to make them from Russia. In 1944, he bought a small New Hampshire distillery, produced a brew consisting of 80 percent potato alcohol and 20 percent straight whiskey, and became a leading distiller of bulk alcohol. Then, as wartime controls on grain alcohol were ending, he acquired some Kentucky distillers and built J. W. Dant into what he called "the crown jewel of Kentucky Bourbons."

As befitting a man of increasing wealth and diverse interests,

Hammer bought a 78-foot yacht, which he often used to make the water section of the voyage between his farm in New Jersey and his office in the Empire State Building. On the farm, he bred Aberdeen Angus cattle and sold bull semen as a profitable hobby.

In 1955, he sold J. W. Dant and Company to Schenley Industries for $6.5 million and retired to Los Angeles with his collection of Old Masters. A year later, he married his third wife (his first two marriages had ended in divorce), a wealthy widow of Irish descent named Frances Hackett. But instead of settling into a comfortable retirement, Hammer made a belated entry into the oil business, with significant consequences for the industry and the American people.

Hammer's debut was no more auspicious than Leon Hess's had been. He was bored and needed a tax shelter. An accountant suggested buying Occidental Petroleum, known as Oxy. Hammer had never even seen an operating oil well, and for an entrepreneur of his vision, Oxy didn't offer much of a challenge. The company was near bankruptcy and its only major asset was nine oil wells south of Los Angeles that produced at most 100 barrels a day.

Hammer declined to buy Oxy for $120,000, but lent the obscure company that sum to drill two wildcat wells in Bakersfield. He could always write the loan off if the wells proved to be dry holes. But both came in and Hammer exercised his stock options to buy Oxy's stock for 18 cents a share on the open market. He now owned a controlling interest in a company with three employees.

After installing himself as president, the astute Hammer realized that he knew nothing about the oil business, so he hired Eugene Reid, who had participated in thousands of drillings in California over four decades. Reid *was* an oil man, respected throughout the industry. Hammer had always had a flair for hiring talent on generous terms, though not all executives chose to remain under his control—in fact, they left his organization in sizable numbers. One of Hammer's favorite mottos is: "Every man has his price."

Gene Reid's first big "play," backed by private financing

Hammer skillfully assembled, didn't come until 1961 in the San Joaquin Valley east of San Francisco. Reid (who died in 1971) didn't find oil, but 600 feet from where Texaco had abandoned an oil well, and 1,000 feet deeper, he struck natural gas.

Oxy, and its president Dr. Armand Hammer, now controlled the Lathrop gas field. It turned out to be the second largest ever found in California. Other strikes followed. Not content with these successes, Hammer diversified. He bought a fertilizer company and the world's biggest fertilizer marketing company, as well as Jefferson Lake Sulphur Company and phosphate deposits in Florida. He made these profitable moves because natural gas is used to make ammonia, which goes into fertilizer, as do sulfur and phosphates. Occidental Petroleum, which was going begging for a buyer at $120,000 in 1956, had sales of $659 million ten years later.

Armed with credibility and cash flow, Hammer was ready to enter the world of intrigue known as international petroleum. Stymied by the power of the majors in the United States, he saw his opportunity in Libya.

In the early 1960s, Libya was a backward North African Moslem land of mostly barren desert where the aging King Idris and his corrupt court ruled 1.7 million desperately poor Arabs. It was also a land where vast deposits of high-quality crude oil had recently been found, exciting the interest of the Seven Sisters, who at that time dominated world production and maintained an orderly world market by dictating both prices and production quotas to the humble producing countries.

Hammer had first visited Libya in 1961, and saw in the desert kingdom a chance to turn his small California-based concern—with no overseas production whatever—into a major international oil company. King Idris and his advisers were not averse to granting a few concessions to this persuasive independent, for any success that he and other independents might have would undercut the power of the unified Seven Sisters to set prices and otherwise call the tune.

In February 1966, Oxy was granted two prime concessions covering nearly 2,000 square miles of bleak desert. Precisely how those concessions were obtained will probably never be known. Critics of Oxy, and disgruntled former executives, have used terms ranging from "compensation to middlemen for assistance" to "overrides" to "bribes." Hammer denies these allegations, but after King Idris was overthrown in 1969, the new Libyan foreign minister declared that Libyan officials had accepted oil bribes and that they were "members of the royal court, who profited thereby on the basis of corruptions, threats and imposing their power."

However the concessions were obtained, Oxy struck oil on them beginning in the autumn of 1966 and continued to do so with phenomenal success, including finding one of the single biggest oil deposits in the world. Once again, Hammer branched out. He used Oxy's suddenly inflated stock to buy more phosphates (which increased in value) and to diversify into coal. In 1968, he paid $150 million for Island Creek Coal, which had huge reserves. Five years later, this subsidiary was earning more than $100 million a year after taxes, and its coal became increasingly valuable as a result of the energy crisis.

In Libya, however, Hammer's realized dream was about to turn into a nightmare. King Idris was deposed on September 1, 1969, by a dozen radical military officers led by the ascetic twenty-seven-year-old Colonel Muammur al-Qadaffi. He and the other members of the Revolutionary Command Council did not play cards, drink alcoholic beverages, or go to nightclubs. They obeyed the Koran and they loathed Israel. They were also convinced that Libya was being exploited by the foreign oil companies and resolved to do something about it.

The only previous attempt by a producing country to curb the oil companies' power had ended in disaster. This was in 1951 in Dr. Mohammed Mossadegh's Iran, which nationalized the Anglo-Iranian Oil Company; as a result, Iran's oil was shunned by the big companies, the Mossadegh government was overthrown, and the Shah was restored to his throne with the aid of the CIA. The Organization of Petroleum Exporting Countries (OPEC) had been formed in 1960 (Libya joined in 1963), but in 1969 it was still an innocuous organization. Shortly after the

Libyan coup, however, the Revolutionary Command Council, acting apart from OPEC, began to pressure the companies for a 20 percent increase in taxes and royalties. There were twenty-one foreign oil companies active in Libya, half of them independents like Marathon Oil, Continental Oil, and Amerada Hess, which operated as a consortium called Oasis; and half of them bigger companies led by Shell, Mobil, and the mighty Esso (now Exxon).

The majors agreed to resist the Revolutionary Command Council. Esso, led by its rugged chairman, Ken Jamieson, an MIT graduate from Medicine Hat, Canada, was particularly hard-nosed. There was a glut of oil at the time, the majors had other sources of supply, notably in the quiescent Middle East, and they didn't want to encourage demands by other producing countries. Rebuffed by the formidable majors, Colonel Qadaffi decided to pick on the independent Dr. Hammer and Oxy.

It was a shrewd if obvious move. Oxy was vulnerable, for, unlike the majors, it was totally dependent on Libyan production, which was the lifeblood of the company and the prime reason for its success. When production was cut from 800,000 to 425,000 barrels a day, Hammer became desperate. Unless he acceded to Qadaffi's demands, which were not entirely unreasonable, the colonel threatened to nationalize Oxy or kick it out of Libya altogether. While negotiating and playing for time, Hammer unsuccessfully tried to get other oil companies to sell him oil at cost as a hedge against losing Libyan production. The majors might well have savored watching the upstart squirm. Finally, in July 1970, Hammer went, more or less cap in hand, to see Kenneth Jamieson.

It must have been an interesting confrontation: the former bargain-basement art dealer and whiskey mogul from the Bronx who hadn't seen a producing well until he was nearly sixty face to face with an engineer who had spent his entire adult life in the oil business and, as chairman of Esso, occupied the office first held by John D. Rockefeller.

Hammer proposed that Jamieson sell him oil at cost; in return, he would stand united with the majors against Qadaffi's price demands. While sympathetic to Hammer's plight, Jamieson refused the deal.

This refusal turned out to be a mistake with significant repercussions for the oil business and the industrialized world. His back to the wall, Hammer gave in to Libya's demands on September 4, 1970, and took the best deal he could get. Of course, he could have been an altruist, stood firm, and watched the destruction of his foreign holdings, but that is not how the oil game is played.

Once the common front had been broken, the other companies began to cave in one by one despite their endless conferences and appeals to the U.S. State Department and the British Foreign Office for help. This was a historic turning point in the oil business because it demonstrated to the other OPEC nations that the oil companies and their governments could not maintain ranks and present a common unyielding front; power had at last been transferred from producing companies to producer countries.

Hereafter, the majors would be whipsawed between further Libyan demands and the demands of the Persian Gulf producers. By the time the October War of 1973 erupted, the major internationals were about ready to accept whatever terms the producer countries dictated. Instead of allocating production among producing countries, they were now reduced to allocating tight supplies to consumer nations. The American epoch in the Middle East oil fields had ended. The once powerful companies that parceled out territories among themselves and dictated how much oil each country could pump and at what price have become OPEC's professional managers with mixed loyalties. Their financial futures—their access to oil—lie with OPEC. But their overall health and very survival lie with the industrialized consumer nations.

The diminishment of the Seven Sisters to handmaidens of OPEC can thus be traced to the entry of independents like Oxy into Libya and the zealous, unyielding pressure applied to them by the Qadaffi government Oxy was able to remain in Libya, although Colonel Qadaffi took over 51 percent of its operations in

August 1973 for a payment of $136 million. But Oxy was not alone in its misfortune. Between 1971 and 1974, Libya nationalized all American and European oil holdings in part or in full.

Despite the U.S. State Department's view that the Libyan government was unorthodox and radical, a sponsor of terrorism, and extremist in its opposition to Israel, U.S. oil companies continued to buy Libyan crude oil. They bought 650,000 barrels a day in 1979, more than one-third of Libyan production, at a cost of $7 billion. There was no American ambassador in Tripoli. Although Colonel Qadaffi denied that his nation was a Soviet base, there were an estimated three thousand Soviet technicians and military advisers in Libya.

On the tenth anniversary of Colonel Qadaffi's coup, a parade, attended by Billy Carter, was held in Benghazi featuring Soviet-made arms, including MIG-25 jet fighters and T-72 tanks never before displayed in the Middle East or Africa. The weapons were paid for with foreign exchange derived from oil royalties; in other words, they were in large part financed by American consumers of Libyan oil.

At times, irked by American and Western European reluctance to take a strong pro-Palestinian stance, the mercurial colonel has threatened to cut off oil exports to the West, which would make the embargo of 1973–1974 and the Iranian shortfall seem like mere inconveniences. It is unlikely that he will carry out his threat, but only because Libya has no other significant source of income. However, on July 1, 1980, Libya, along with Nigeria and Algeria, raised the price of crude oil to $37 a barrel, the highest price among OPEC members. On December 31, 1978, Libyan crude had cost $13.85 a barrel.

The ripple effect of the $37 price resulted in further rises in the retail price of gasoline, home heating oil, and other refined petroleum products in the United States and around the world. But Occidental Petroleum continues to prosper. In 1979, it was the eleventh largest American oil company with earnings of $561.6 million on sales of $9.6 billion.

While remaining in active control of Oxy, Hammer somehow found time for other business ventures. In 1972, he returned to

Russia with a large retinue, including Sargent Shriver as his lawyer. When in 1964 Hammer had tried to put together a fertilizer deal with Anastas Mikoyan and Khrushchev, it hadn't come off. Now, in the era of détente following the Nixon-Brezhnev summit meeting, prospects seemed brighter for making money in Soviet-American trade. Hammer met with eighteen senior ministers in five days and discussed a host of possible deals. On his next trip, he carried a letter from Nixon praising his efforts, met with General Secretary Leonid Brezhnev and Premier Kosygin, and gave copies of one of his favorite books, *Dr. Atkins' Diet Revolution,* to certain overweight Soviet officials. The Russians seemed to take him seriously. He had done business there before and he had, after all, once conferred with Lenin himself.

On future regular trips, Hammer proposed a fertilizer deal, a natural gas deal, a nickel- and metal-plating deal, a trade center deal. After endless rounds of negotiations, at the end of June 1974 he signed contracts for a complicated $20 billion fertilizer deal spread over twenty years. Barter was at its heart: Oxy would send phosphates and technology to the Russians in exchange for potash, urea, and ammonia. Hammer is by far the largest American businessman in the Russian trade. He occupies a lavish Moscow apartment for about a total of one month each year. From its balcony, he has a beautiful view of the Kremlin.

Not that Dr. Hammer hasn't encountered a few setbacks. He pleaded guilty in the federal district court in Los Angeles to making an illegal contribution of $54,000 to the 1972 Nixon campaign. An Oxy subsidiary, the Hooker Chemicals and Plastics Corporation, was charged with dumping toxic wastes into the Love Canal in Niagara Falls, New York, and as a result suffered losses and had to defend itself against costly lawsuits over waste disposal. After the Soviet invasion of Afghanistan in late 1979, President Carter brought trade with the Russians to an abrupt halt and Dr. Hammer's twenty-year $20 billion fertilizer deal with the U.S.S.R. also came to a halt, at least temporarily. Finally, in July 1980, the Securities and Exchange Commission charged that Oxy had failed to disclose hundreds of millions of dollars' worth of potential liabilities arising from its environmental practices, including those at Love Canal, and ordered the company to establish the most extensive monitoring program ever mandated by a federal agency.

That August, the eighty-two-year-old Dr. Hammer named A. Robert Abboud, former chairman of the First Chicago Corporation, as Oxy's new president and chief executive officer. Abboud was the fifth potential successor appointed by Dr. Hammer since 1968, and this game of musical chairs did nothing to build up investor confidence in Oxy. The chairmen of the Seven Sisters are pensioned off at age sixty-five, but since Hammer owns the controlling interest in Oxy, no one can force him to retire. And he may wonder why he should, when there are more deals to pull off.

Dr. Hammer turned up in August 1980 in Islamabad, Pakistan, to sign documents relating to a joint-concern deep-drilling oil project in which Oxy will invest $17 million. But Oxy's biggest deal and greatest gamble is in shale oil, on which the company has spent more than $100 million of its own money and $30 million in federal funds.

Shale is a hard rock containing kerogen, which yields a low-grade form of crude when heated. This waxy substance has been known about for a century, but the oil companies showed no interest in developing it while conventional domestic oil was in ample supply and cheap to produce. In 1980, however, spurred by an average OPEC price of $32 a barrel, tax credits, loan guarantees, and other financial supports offered by a federal government anxious to encourage alternate fuels, the companies were ready to give this perennial pretender a chance.

Dr. Hammer, the industry's most fervent champion of shale, is convinced that Oxy could profitably produce oil from shale for $25 a barrel, with commercial operations beginning by the mid-1980s and reaching full production by 1988. He also believes that Occidental Oil Shale Inc., a subsidiary, with its in situ (in place) extraction process, can avoid residue disposal problems and drawing off undue amounts of water from parched western lands at its experimental plant in Logan Wash, Colorado.

There are those who disagree with Dr. Hammer. "A lot of spear shaking and saber rattling goes on about how shale is now clearly economical," Roger Loper, president of the Chevron Oil Shale Company, a unit of Socal, has said. "It's absolute balderdash. Nobody has proven we can do this, and a lot of blood, sweat and tears have to come down the river and go over the dam before we can be sure." According to Philip Robinson of the U.S. Office of Technology Assessment, shale oil would have to sell for an average of $62 a barrel (in 1979 dollars) for the next twenty-two years to provide a 15 percent return on investment in a 50,000-barrel-a-day plant costing $1.7 billion.

The stakes are enormous, and in dispute. Shale advocates estimate that perhaps 400 billion barrels of oil might be recoverable; this is more than double the proven reserves of Saudi Arabia. But Philip Robinson concludes that existing environmental laws will limit shale oil production to between 1 and 1.5 million barrels a day.

Dr. Hammer clearly wants to lead the way in transforming shale oil into a major factor in the American energy picture. It would be the biggest deal ever brought off by this most unlikely of oil men.

The Gambling Oil Man

Leon Hess and Dr. Armand Hammer, each in his own unconventional way, have made it into the big time of the oil game. Another kind of oil man who is still trying to make it is John Patrick Gallagher.

"Smilin' Jack" Gallagher, as he is known, is chairman and chief executive officer of Dome Petroleum Ltd., an oil and gas company headquartered in Calgary, a booming city in Alberta. This sparsely populated western province whose southern section slams into the Rocky Mountains holds 85 percent of Canada's proven oil and gas reserves, half of its coal, and huge deposits of tar sands.

Founded in 1876 as a fort for the Northwest Mounted Police, then a prairie cow town best known for its mixture of beer and tomato juice known as Calgary Redeye, this fastest growing large city in Canada now has more than half a million residents. It is host to over a score of foreign banks and its modern skyline is alive with new structures bearing names like Sun Oil Building, Mobil Tower, Shell Center, BP House, Esso Plaza, Gulf-Canada Building, and Dome Petroleum Tower.

There is no mystery about why Alberta is booming, with its wealth of natural resources, unpolluted air, and low taxes. Since the provinces own their own mineral rights under Canadian law, Alberta can charge Ottawa and the consuming provinces of the east what the market will bear for oil. Its leaders have not hesitated to do so, earning them the sobriquet of "the Blue-Eyed Sheiks," as in the title of a book published in 1979 by Peter Foster.

Dome has done well in western Canada, where 60 percent of its revenues come from natural gas liquids, and where it owns 4,000 miles of pipelines, part of a gold-mining concern, and has discovered some respectable oil wells. Although it doesn't have the financial resources of the Seven Sisters, Occidental Petroleum, or Amerada Hess, Dome's stock is traded on the major Canadian and American exchanges and has attracted American institutional investors like the endowment funds of Princeton, MIT and Harvard. But it is not in Alberta that "Smilin' Jack" Gallagher hopes to make it into the big time. It is more than 1,500 miles north of Calgary in the Beaufort Sea, which is part of the Arctic Ocean between northern Alaska and Canada's Arctic archipelago. Vast oil deposits have already been found 400 miles to the west in Alaska's Prudhoe Bay.

Dome holds about one-third of the ice-choked Canadian Beaufort's leased acreage. A gamble there is not for the fainthearted, but many experts believe that the geological formations are favorable—a river delta of the type that has produced large oil fields elsewhere and structures that could contain large deposits. Finding these deposits and extracting them from the Arctic ice in subzero temperatures is another matter. Still, Gallagher is optimistic. "We think of the Beaufort as the greatest potential, unexplored oil and gas basin in the world," he says. "All the great oil fields being found now are in the deltas of major rivers, the Mississippi, the Orinoco in Venezuela, the Niger Delta, the Euphrates in the Persian Gulf. In the Beaufort there's the Mackenzie River. We feel very positive that it's going to be a very productive oil and gas basin."

Gallagher's career began in the 1930s when, as a geology student from Winnipeg, he earned $2.50 a day helping to map Canada's northern wildernesses. Then he spent thirteen years in the Middle East, Africa, and South America as a geologist for Shell and Esso. During World War II, he helped deactivate Nazi acoustical mines air-dropped into the Suez Canal, and in Ecuador found Japanese supply caches hidden for an anticipated assault on the Panama Canal. "It was great," Gallagher says. "We made five trips across the Andes by foot; had to cut our own way. I spent 23 out of 24 months in the jungle in Ecuador. The area had good potential, but politically it wasn't ready. That's the problem with foreign work. It's O.K. for the majors to be in 20 or 30 countries and maybe two of them pay off. But it's too big a risk for the independents. That's why Dome has essentially stayed out of foreign work."

Dome was founded in 1950, at a time when the majors had already wrapped up the most promising exploration leases. So Dome surveyed the hostile Beaufort Sea, found potential oil-bearing structures, and acquired drilling rights. The problem was to find backers for a costly project in the Arctic that could turn out to be a bust.

Here Gallagher had to be as nimble as he was in the Ecuadorian jungle. First, more cautious companies with drilling rights turned to Dome to do the actual work. Gallagher took payment in cash and in partial rights to whatever was discovered, which broadened Dome's holdings. Then he dealt off most of the financial risks to others by farming out part interests in Dome's holdings to gas companies and financiers, a form of equity financing that uses land rather than shares. Next he urged the Canadian government to grant increased tax incentives to attract exploration money. The government, interested in reducing Canada's dependence on imports, responded by tacking a tax shelter onto the October 1977 federal budget. Some have dubbed it the "super writeoff" or the "Dome budget."

Dome operated four drill ships that could drill to depths of 12,000 feet and cost $10,000 a half hour to use, and an icebreaker, *Canadian Kigoriak*, on a 1979 operating budget of about $160 million (Canadian dollars). The weather permits operations only in the summer months, or for about 160 days a year. In October 1979, Dome announced that it had struck oil at its Kopanoar-13 well that could be capable of producing more than 12,000 barrels a day. Dome owns a roughly 48 percent interest in the discovery well; the other owners are Gulf Canada (25 percent); Hunt International (25 percent); and Columbia Gas Systems (1.875 percent). Not even Smilin' Jack Gallagher would forecast that his company's original goal of delivering sizable quantities of oil from the Beaufort Sea by tanker in 1985 can be met without further drilling and evaluation of Kopanoar-13 and other wells in the area with such names as Tarsiut A-25, Orvilruk O-3, and Nerlerk M-98.

Gallagher's gamble illustrates a cardinal fact about the oil game today. Any major discoveries in North America are likely to take place either off the coastlines of highly populated areas or in remote, hostile frontiers like the Beaufort Sea. They will be high-risk operations requiring complex financing deals and tax incentives. And before any discoveries can be expected to provide relief from the energy crisis, ways will have to be found to transport the oil to markets at reasonable cost and minimal damage to the environment.

John Patrick Gallagher is not daunted by these obstacles. Up in his office in Calgary, which Canadian writer Mordecai Richler says "is going to be a helluva city when they get it uncrated," Gallagher undoubtedly sees things differently from those who occupy the executive suites of Big Oil in Los Angeles, New York, Houston, and Chicago.

A Very Big Oil Man

If one had to select a man to personify those who run Big Oil today, John Elred Swearingen would certainly be in the forefront of the candidates. He is a large man with wavy gray hair who speaks softly and puffs regularly on a pipe. In 1980, he was chairman of the board and chief executive officer of Standard Oil Company (Indiana), America's sixth largest oil company and a direct descendant of the old Standard Oil trust, marketing under the Amoco brand. The year before, he had been board chairman of the American Petroleum Institute, the industry association to which all the majors and leading independents belong, and thus the chief spokesman for the industry.

Swearingen's career follows the pattern for leaders of the industry. He has worked in no other business but oil, and for no other company but Indiana Standard. He was born in Columbia, South Carolina, in 1918, has a B.S. degree from the University of South Carolina and an M.S. from Carnegie-Mellon, and is a chemical engineer who wears a Phi Beta Kappa key. He joined Indiana Standard's research department in 1939 and worked there until 1947. Then he rose steadily: first, general manager of production, then vice president of production, executive vice president, president, and finally, chairman of the board in 1965.

Indiana Standard was something of a weak sister when Swearingen took it over. He described it as a "large, mediocre company," and said: "If nothing had been done, I'm sure the company would have lumbered along as it had been doing, but it probably would have declined in relative terms."

What Swearingen did could not have endeared him to some employees of the lumbering company. Management deadwood was cleared out and thousands of workers were fired as refineries were modernized and operations tightened. Then Swearingen greatly expanded exploration and production, the most profitable segment of the oil business.

Amoco Production Company drilled nearly one thousand wells

in the United States in 1978, about three hundred more than the next most active company. It has discovered oil and gas fields in the Overthrust Belt in the Rocky Mountains; two areas in western Alberta; and in the Tuscaloosa Trend in southern Louisiana. It holds leases on 34 million acres in the United States, mostly onshore, more than twice that of any other company, in line with a decision made in the early 1970s to concentrate on exploration in the United States. However, Swearingen has been cautious about joining the high rollers in their bids for U.S. offshore acreage. In view of other companies' failures drilling in the publicized Baltimore Canyon off the East Coast, his refusal to pay what he calls "outlandish" bonuses for offshore rights may prove to be justified. Swearingen is obviously betting that, with deeper drilling capabilities and new tools, there are still sizable deposits of oil and gas to be found in Canada and the United States. With the phasing out of price controls on U.S. crude oil, Indiana Standard is in the best position of all the big companies to profit from the reviving search for oil and gas in the United States.

Like all the leaders of Big Oil, Swearingen has had to take big disappointments in stride. Indiana Standard bought acreage on the North Slope of Alaska, but found no oil there. A venture into copper mining in Zaire produced heavy losses because of political upheavals. Indiana Standard's production of 200,000 barrels a day in Iran was shut down after the fall of the Shah.

Another of Swearingen's few failures involved Dr. Armand Hammer. They met, at Swearingen's request, in Hammer's office near the UCLA campus one afternoon in November 1974. Hammer was suspicious because he knew that Swearingen, like Ken Jamieson, considered him an outsider. After hours of listening to this old man with the leathery skin and hooded eyes talk about his Soviet deals, his art collection, and his troubles in building Oxy into a major oil company—to all of which Swearingen attended with uncharacteristic patience—the chairman of the board of Indiana Standard came to the point. He had a deal to propose—as it turned out, the last one Hammer wanted to hear about. Indiana Standard, attracted by Oxy's Libyan crude and the huge reserves of its Island Creek Coal subsidiary, wanted to take Oxy over. The merger would be the biggest in the history of U.S. business, resulting in a company larger than U.S. Steel or IBM. Since there wouldn't be room at the top for two strong-willed characters like Swearingen and Hammer, it would mean the forced retirement of the doctor.

Even though Hammer was seventy-six years old at the time and had all the money any man could ever need, he had no intention of giving up active control of his creation. He told Swearingen that he wasn't interested in this particular deal and said that he hadn't changed his mind when Swearingen telephoned him from his hotel later that evening. "I will," Hammer told Swearingen, "fight you all the way down the line." Hammer was determined and the stage was set for the biggest legal battle ever to involve a forced corporate takeover. Just two months later, however, Swearingen announced that Indiana Standard was abandoning its bid to acquire Oxy, undoubtedly out of fear of antitrust action.

Even without Oxy, Indiana Standard continued to prosper. Its marketing arm, Amoco Oil Company, is second only to Shell in gasoline sales in the United States and it is first in the profitable lead-free market. In 1979, Indiana Standard posted a net income of $1,506,618,000 on sales of $18,610,347,000.

Swearingen directs Indiana Standard from Chicago's second tallest building, the company's eighty-story marble tower east of Michigan Avenue by the lake. On a salary of $635,000 a year, he lives in a sixteenth-floor three-bedroom apartment on Lake Shore Drive overlooking Lake Michigan with his wife Bonnie, and maintains a home in Palm Springs, California. Bonnie Swearingen, who is some fifteen years her husband's junior, had a short acting career in Hollywood in the 1950s before marrying and divorcing two Texas oil men. She married John Swearingen in 1969 after he divorced his wife of more than twenty years (he has three married daughters from that first marriage). According to a reporter writing in the *New York Times* of August 25, 1980: "John and Bonnie Swearingen of Chicago may well be the most colorful pair in the upper echelons of corporate America."

They certainly are a study in contrasts. She is a high-living former Miss Alabama, blond, brash, and sexy in the curvy 1950s style. Outspoken in her taste for champagne, emeralds, and "the smell of oil, which should be bottled like perfume," Bonnie Swearingen is capable of telling an interviewer with a straight face that "I dreamt every night of Napoleon before I met John."

The Swearingens attended the Inaugural Gala in January 1981. Bonnie Swearingen wore a satin ball gown and an emerald-and-diamond necklace with matching earrings. "All of the women here have husbands who have worked hard for what they have," she observed. "What have we worked for if we can't enjoy it? It's getting a little tiresome to always have to apologize for ourselves."

Pointing to her bodice, she continued, "If a little girl from Alabama whose father was a minister can appear in public wearing beautiful jewels and gowns it should be a symbol to everyone that they can do it, too."

John Swearingen has certainly worked hard for what he has. He is known as a blunt, impatient, overbearing, even dictatorial chief executive (like Napoleon?), but also as one who delegates authority, gives credit to subordinates, and disdains yes-men. In his role as board chairman of the American Petroleum Institute in 1978 and 1979, he became a fixture of public forums and TV talk shows, given to abrasive comments about "hysterical oil critics" and the "naive Department of Energy." Undoubtedly, he reflected the opinions of his fellow leaders of Big Oil, although few of them—at least in public—proclaimed them so bluntly.

In September 1979, Swearingen, who travels constantly and is away from Chicago about half the year, flew to Rumania to attend the tenth World Petroleum Congress—which meets every four years to exchange technical information—along with experts from more than seventy countries. From the Imperial Suite at Bucharest's leading hotel, he told American reporters that private American oil companies could supply the American people with all the oil and gas they needed—but only if the federal government would stand back and leave the industry alone.

"I think we've got a bunch of amateurs running the government. And their primary concern is the campaign of 1980," Swearingen said. "Mr. Carter is president of the United States, but I don't believe he's acknowledged as leader of the Democratic party. I don't believe his energy bill is going to be enacted in the way he proposed it. Mr. Carter is making an attack on the oil industry one of the principal components of his campaign for reelection."

Concerning Charles Duncan, who had just replaced James Schlesinger as energy secretary, Swearingen allowed that their first meeting had been very friendly, but said it wasn't clear how much control Duncan would have over energy policy. "After all, he's not an oil man, and he doesn't even know the jargon of the business," Swearingen said. "He's a good businessman, but he has a lot of catching up to do."

Swearingen complained that President Carter had ignored the advice of the oil industry in formulating his energy policies. "It's a one-way street," he maintained. "By that I mean I can write letters and get no replies. And the president has not seen fit to call in people from the oil industry and seek their advice."

Swearingen then advised the reporters gathered in the Imperial Suite that "the public is being confused by all this blather about oil company profits and oil companies holding back production." He maintained that the world had sufficient oil and gas to last at least for one more century but, he cautioned, "not at yesterday's prices, not at today's prices."

John Swearingen may not be right about at least a century of supplies, but he's certainly right about the inevitability of higher prices. And whether his comments about government are dismissed as paranoia or praised as a blunt telling of the truth, they do provide an insight into the thinking of the men who run Big Oil.

Breaking into the Big Leagues

In 1963, the Atlantic Refining Company of Philadelphia bought the New Mexico–based Hondo Oil and Gas Company for stock. Joining Atlantic as a director was Robert O. Anderson, owner of Hondo, who examined Atlantic's unimpressive earnings and determined that the management of this regional, medium-sized refiner was too conservative. Anderson got rid of the chairman and several key officers, took control himself, and began looking for mergers.

He had unsuccessful talks with Pure Oil Company, Sunray DX, and Union Oil Company before concluding a merger in 1966 with Richfield Oil Corporation, number four in West Coast marketing. The timing couldn't have been better. Richfield stock was held by Cities Service Company and Sinclair Oil Corporation, and the Justice Department had ordered them to divest it. Richfield was also short of cash, having just outbid Atlantic for some lease acreage on Alaska's North Slope. Three years later, in a $1.6 billion deal, Atlantic Richfield (Arco) acquired Sinclair, and along with it the basic component of its pipeline system, some additional North Slope acreage, and Sinclair's large Houston refinery. But the move that put Arco into the big leagues turned out to be that acquisition of Richfield's North Slope acreage. In 1968, Arco struck oil at Prudhoe Bay, which at 10 billion barrels is the largest domestic discovery on record (part owners are Exxon and Standard Oil of Ohio). In 1980, the field produced more than 1.5 million barrels a day, some 16 percent of U.S. production.

Before that oil could be brought to market, Arco had to go into debt (some $4.7 million) to finance its share of the Trans-Alaskan pipeline, the largest privately financed industrial project in history. But Anderson has never been afraid to spend money to make money. He built the most modern refinery in the United States at Cherry Point, Washington, to handle the high-sulfur Alaskan crude. He has diversified into petrochemicals, coal, and solar energy, and he bought the Anaconda Copper Company in

1977 (the Federal Trade Commission forced Arco to divest itself of some of its copper interests in 1980). But oil, domestic oil, is still Anderson's principal game. In 1980, Arco got 95 percent of its 568,000-barrels-a-day production from U.S. sources. That year the company scheduled a bold five-year $20 billion capital-spending program, 60 percent of which is earmarked for prospecting and developing U.S. oil fields. In 1979, Arco was the seventh largest U.S. oil company, with sales of $16,233,959,000 and net income of $1,165,894,000.

Anderson, an energetic silver-haired man whose father was a prominent Chicago banker, moved Arco from the conservative surroundings of Philadelphia to New York City and then to Los Angeles. He only spends about 60 percent of his time on the company. Unlike most oil men, he has other interests. Land, for one. He is the largest individual landowner in the United States with more than 1 million acres on eleven ranches in the Big Bend country of Texas and New Mexico. He has one of the biggest commercial cattle and sheep operations in the country, and breeds quarterhorses, thoroughbreds, and Arabians. Before the Iranian revolution, he owned vast tracts of farmland in Iran. With David Rockefeller and a Brazilian partner, he also owned a million-acre ranch in the Brazilian state of Mato Grosso, but sold it early in 1980.

In that year, Anderson and his wife of forty-one years, Barbara (they have two sons and five daughters), spent much of their time at their 150,000-acre Circle Diamond cattle ranch, complete with polo field, in the Hondo Valley in southeastern New Mexico. He commutes to Arco's Los Angeles headquarters from New Mexico for the first two days of each week in a company plane, staying in an apartment in Los Angeles for which Arco pays $21,000 a year. He is a major contributor to the Republican party, a leading collector of Indian art, and a sponsor of research into numerous global issues, most notably through the Aspen Institute for Humanistic Studies.

In addition to the Circle Diamond ranchhouse, Anderson maintains homes in Roswell, New Mexico; Aspen, Colorado; and Durango, Mexico. He seems able to enjoy the rewards his entrepreneurial talents have brought him, whether these take the form of grouse shooting in Scotland, restoring an old New Mexico town, owning America's oldest maker of boots and other Western wear, or speculating about the fate of the earth's nonrenewable resources. Nobody knows what his income is from his various private businesses, but in 1979, his salary, bonus, and incentive fees from Arco came to more than $1.2 million and he and his wife owned more than 220,000 shares of Arco stock worth some $10.3 million. All of this started in the late 1930s with a $50,000 investment in Hondo Oil.

Chairmen of the board like Anderson, Hess, Hammer, Gallagher, and Swearingen are also the chief executive officers of their companies and as such run the show. Not to be overlooked (although they usually are by the public) are the presidents who rank just below them and are normally destined to succeed them.

The president of Arco, Thornton Bradshaw, is about as far from the stereotypical oil man as it is possible to get. He has a B.A., an M.B.A., and a doctorate in Commercial Science from Harvard, where he once taught at the Business School. In 1980, when he was sixty-two, Bradshaw was an overseer at Harvard and a director of the Conservation Foundation. He has been quoted as saying that "the whole energy situation is too important to be left to the oil and energy industries, the same as war is too important to be left to the generals."

In a speech to Massachusetts Institute of Technology alumni in June 1980, Bradshaw, citing "the tired old oilfields of the United States," warned that the country's oil production was slipping at a faster-than-predicted rate and said: "We must face the fact that our real energy policy through the 1980s will be dependency on Middle Eastern oil." He urged that U.S. foreign policy address itself to the problems of that volatile area and to finding a solution to the question of a Palestinian homeland.

Like John Swearingen, Bradshaw is against government intervention into the operations of the domestic oil industry, but he couches his resistance in more diplomatic language. Arco's 1977 annual report informed the company's stockholders:

> Industry needs to be more open to change, to communicate better with its many constituencies both within and outside government. Companies like Atlantic Richfield should not, in Dylan Thomas' words, go gently into greater regulation, but we need to make clear why it is in everyone's interests—and not just ours—to resist needless government interference.

In contrast to the industry's hardliners, Bradshaw is not against the windfall profits tax. "Decontrol is a bitter pill for the American people to swallow, and the excess profits tax is a necessary sugar coating for that pill," he has said. "Atlantic Richfield never fought against it."

Bradshaw would recognize what he calls "the very legitimate claims of environmentalists," while speeding the settlement of environmental disputes that block construction of energy-producing projects. He also advocates speeding up the development of synthetic energy, although he thinks that federally aided projects might be able to produce 500,000 barrels of oil a day by 1990 rather than the 2.5 million barrels the government has envisioned. Without serious conservation moves, including standby rationing, Bradshaw fears that America will be confronted with a choice between shortages and increased imports in the early 1980s. "When we think of alternatives to oil in the 1980s, we are simply stuck with coal and nuclear," he has observed.

While they are as concerned about profits as any other oil men, Anderson and Bradshaw have taken action in that vague area known as "corporate responsibility." Among other projects that might be dismissed as quixotic by some in the industry, Arco has engaged in an aquaculture experiment in the Santa Barbara Channel designed to increase the abalone population there, paid for thousands of trees planted by schoolchildren in Los Angeles and Ventura counties, and taught women the basics of automotive maintenance, free of charge, at its participating service stations. Arco donates millions of dollars each year to the Atlantic Richfield Foundation, which in turn distributes funds to various worthy educational, social, and environmental projects. In 1980, the Atlantic Richfield Foundation joined with the MacArthur

Foundation to rescue *Harper's*, the nation's oldest monthly magazine, from its announced demise.

But now the oil game will lose the man whose outspoken concern for social responsibility frequently cast him as a maverick among fellow oil executives. Since 1972, Bradshaw had been an outside director on the fifteen-member board of directors of the RCA Corporation. After suffering a series of management fiascos, and absorbing a precipitous three-year drop in the profits of its NBC subsidiary (which continued to trail CBS and ABC among the TV networks), the diversified broadcasting and electronics giant announced in February 1981 that its chairman and chief executive officer, Edgar Griffiths, was stepping down. He will be succeeded by Thornton Bradshaw.

Bradshaw had been planning to retire from Arco in two years in any case and had groomed his successor. With Bradshaw's departure from America's seventh-largest oil concern for RCA, the executive suites of Big Oil will revert to business—and profits—as usual. For certainly no one else among the men who run the show has often ignored industry dogma or engaged in such activities as serving on the boards of the Los Angeles Philharmonic and the Aspen Institute for Humanistic Studies; turning down an offer to become president of the University of Southern California; and running many of the business affairs of the *Observer* (Britain's oldest Sunday newspaper, bought by Arco in 1976 and sold in 1981). In the introduction to a book he coedited entitled *Corporations and Their Critics*, Bradshaw wrote "Those who believe, as I do, in the intrinsic value of the decentralized market system must act now to develop a more humanistic, responsible form of capitalism to meet society's demands as well as satisfying its needs."

Pronounced Tav-uh-la-REE-uss

None of the executives described thus far belongs to the world of the international giants, the Seven Sisters. One who does is the president of Mobil, and he is as different in approach from Thornton Bradshaw, Harvard '40, as Bradshaw is from the good ol' boys encountered at the Houston Petroleum Club.

William P. Tavoulareas, the Brooklyn-born son of a Greek immigrant, became president of America's second largest oil company in 1969. He is not an oil man. He is a numbers man, a tough, skilled negotiator and deal maker whose tactics have sometimes alienated the engineers and geologists who make up the fraternity that runs most of the other majors.

Tavoulareas served Mobil for twelve years as an accountant and lawyer before becoming the first head of the company's planning department in 1959. "Tav," as he is known, sometimes employs barnyard language and gets entangled in his fractured syntax. He sets much of Mobil's policy, but defers in the final analysis to the company's chairman, Rawleigh Warner, Jr., a Princeton graduate who is greatly concerned with Mobil's public image. Sometimes they seem like the good guy–bad guy duo in old detective movies.

Warner is aware of Tav's hard-nosed reputation with the competition and once told a reporter from the *Wall Street Journal:* "Some of them don't trust him, just because he's so able. He's . . ." Warner's observation was interrupted by Herbert Schmertz, Mobil's vice president for public affairs: "They don't trust *themselves.*" Warner continued: "I believe I should not interfere with people who I believe are doing an extraordinarily good job. And I think Tav is as adroit a negotiator as I've ever seen."

Mobil's profits would seem to bear out Warner's estimate, and Tavoulareas believes that Mobil has a right to those huge profits. "We say the threat against the industry and the capitalist system is a real thing," he told the same *Wall Street Journal* reporter. "You've gotta start with that. We've gotta get up and step up and defend ourselves."

Mobil has certainly done that, to the tune of a public relations budget of $21 million a year, the largest in the industry. But Mobil has a few problems that it doesn't discuss in its op-ed page ads. It has lost a number of its valuable exploration experts to other companies; in 1978, it brought a lawsuit (later dropped) against Superior Oil for luring away thirty of them. Compared to its fellow giants, Mobil has a relatively weak position in domestic oil reserves, which means it is forced to obtain about 53 percent of its crude from less secure foreign sources. So it was not surprising when, in the summer of 1979, Mobil broke ranks with its fellow majors and took the judicious position that it would not object to the retention of price controls on existing oil (i.e., oil that Mobil hasn't much of) provided they were lifted on newly discovered oil without the imposition of a windfall profits tax (i.e., oil that Mobil is trying to find). Needless to say, this proposal did not endear Mobil to majors with substantial holdings of existing domestic oil, and its vociferous opposition to the windfall profits tax led President Carter to call Mobil "perhaps the most irresponsible company in America" (Mobil employees began appearing for work wearing buttons reading: "Call Me Irresponsible").

Meanwhile, Tav leads the aggressive search for new supplies and acquisitions, by all accounts brilliantly. He does a lot of dealing in the Middle East. At the time of the oil embargo, Mobil managed to increase its share in Aramco from 10 to 15 percent, costing its partners—Exxon, Texaco, and Socal—hundreds of millions of barrels of oil.

When Tav was asked by the *Wall Street Journal* reporter if he enjoys playing the heavy in business negotiations, he replied: "Only when I'm dealing with a wise-ass."

Even the Big Boys Have Problems

Much more typical of the men who run the Seven Sisters is John K. McKinley, the new Texaco chairman. McKinley is an oil man *par excellence* whose favorite saying is: "If it ain't broke, don't fix it," delivered in a faint southern accent. He became chairman of the board of America's third largest oil company (after Exxon and Mobil), at the age of sixty on November 1, 1980, succeeding Maurice F. Granville. In that job, he has to cope with a number of things that need fixing, for Texaco is generally regarded as the weakest of the Seven Sisters.

McKinley has worked in no business but oil and for no company but Texaco. He was born in Tuscaloosa, home of the University of Alabama, where his father was a professor of education. He grew up on campus and enjoyed watching the Crimson Tide win football games and listening to classical music. McKinley got a B.S. in chemical engineering and an M.S. in chemistry from the University of Alabama and in 1941 went to work at Texaco's vast Port Arthur, Texas, refinery. Three months later, he was in the U.S. Army, where he rose to the rank of major in a field artillery unit in the European theater. After World War II, he rejoined Texaco and began his steady climb to the top through various managerial positions.

By 1967, McKinley was vice president–petrochemicals. In 1970, he was put in charge of supply and distribution, which made him responsible for some 160 Texaco tankers linking the approximately 130 countries the company operates in and for getting crude from wells to refineries to customers on tight schedules. Later, he was made senior vice president in charge of worldwide refining, supply and distribution, petrochemicals, and research. He holds several patents for chemical and petroleum processing, among them one for a gasoline additive that prevents engine corrosion. In 1971, he was elected a director and president of Texaco. Now, as chairman and chief executive officer, McKinley operates out of Texaco's headquarters in Harrison, Westchester County, New York. A slim, leathery man of courtly manner with a pleasant sense of humor, McKinley's face can take on the hard look of a Mississippi riverboat gambler when it comes to business.

For those who wonder what the leaders of Big Oil do, McKinley's functions are instructive. He is the final authority in an organization of some sixty-six thousand employees worldwide. Management is decentralized along geographic lines. The company's various functions are broken down and grouped into a number of distinct integrated oil companies. The executives responsible for Texaco U.S.A., Texaco Canada Inc., Texaco

Europe, Texaco Latin/America, Texaco Chemical Company, and Texaco International Exploration Company all report ultimately to John McKinley. He checks to see how profitable their individual operations are and what, if any, changes are required. In turn, Texaco's 415,300 shareholders and fifteen-person board of directors check to see what kind of bottom line McKinley produces, for that is the name of the game.

Right now McKinley and his management team face some serious problems. Consumers who were outraged at Texaco's enormous earnings in 1979 (they rose 106 percent over 1978, to $1.76 billion on revenues of $39.1 billion) might be surprised to learn that in the oil game, Texaco's headquarters are known as "Port Fumble," a sobriquet earned during a string of lean years in which its growth rate in earnings per share was the lowest among the major oil producers.

The years between 1956 and 1965, when the brilliant, imaginative Augustus C. "Gus" Long was running the show, were good years for Texaco. After he retired, trouble set in, and when his successor resigned, Long was prevailed upon to return as chairman from 1970 to 1972.

Texaco's stock traded above 45 in the 1960s, but hit a 1970s low of 15 (it climbed back to a 1980 high of 54⅜). In the glory Gus Long era, Texaco had plenty of U.S. reserves and took pride in being the only oil company that marketed gasoline in every state. It also carried cheap crude from Saudi Arabia to Japan and Europe.

Then, in October 1973, OPEC took supply and pricing control away from the Seven Sisters, precisely when Texaco's domestic reserves were beginning to dwindle. The company missed out on the choicest North Sea tracts and its effort to find oil in Alaska ended in failure. It was caught with forty thousand U.S. gasoline stations, many of them small and inefficient, in the self-service marketing revolution and had to close ten thousand of them and back out of all or part of ten states.

It is John McKinley's task to streamline operations, improve efficiency, and stem these reverses. There have been some encouraging signs. Texaco was the first company to find natural gas in the Baltimore Canyon in the Atlantic Ocean off New Jersey. It is a leading marketer of gasohol, a blend of 90 percent lead-free Texaco gasoline and 10 percent ethanol made from agricultural products. It has signed an agreement with Southern California Edison to build a $300 million test facility for coal gasification along with other investors.

Texaco has also formed a six-year partnership with Mesa Petroleum Company, a leading independent producer, to explore for oil and gas on some 1.9 million acres of U.S. leases held by Mesa. Texaco will put some of its recent profits into an adventurous $6.6 billion exploration program in the United States through 1985. It is hoping finally to get some production from the North Sea with the start-up of its Tartan field there. And it eventually expects to make some money out of its holdings in coal, shale, uranium, and tar sands and experiments with synthetic fuels.

John McKinley, for all the support and advice he gets from a small army of presidents, senior vice presidents, and vice presidents, bears the ultimate responsibility for the outcome of all these diverse projects. He works hard and lives quietly in the Connecticut suburb of Darien with his wife, whom he married in 1946. They have two sons. He takes the familiar industry line that big profits are as good for America as they are for Texaco, for they make possible expanded domestic exploration. He champions the workings of the free market over restrictive government regulation, although in more temperate language than that used by the old-time east Texas independent oil man and self-styled Jeffersonian Democrat J. R. Parten, who has been quoted as saying: "Get the government staff off the people's backs"—and, by extension, out of the economy.

In shepherding a Texaco comeback, McKinley has to hope that the company's supplies from the Middle East, Indonesia, and Nigeria are not cut off by political upheavals. Any such cutoffs of lengthy duration would spell disastrous shortages for the United States and would come close to breaking Texaco to an extent that even McKinley couldn't fix.

The Committee

McKinley's management style differs from that of independents like Hess and Hammer, who own a controlling interest in the corporation they run and can call the tune without looking over their shoulders. Rather it resembles the style of the men who head the other multinationals. Exxon, for example, is run by a management committee headed by chairman Clifton Garvin, Jr., and includes the president of Exxon, Howard Kauffmann, as well as six senior vice presidents. All eight, like McKinley, rose slowly through the ranks to their present eminence. Ultimate power, as in most corporations, lies with the board of directors. At Exxon, the board consists of the eight members of the management committee, who are basically high-level employees, and eleven outside directors. Reporting to the management committee on the status of their operations and for decisions on their future plans are thirteen autonomous operating companies, or affiliates, and seventeen staff departments involved in such activities as law, corporate planning, and public affairs.

Seated like eight proconsuls in swivel chairs around a long tapered table fifty-one floors above the street in Rockefeller Center, the management committee oversees activities and decides on expenditures that will return a suitable profit for a worldwide corporation whose 1979 revenues of $84.8 billion almost matched the gross national product of Spain. But none of these eight men, including Garvin himself, are indispensable. They are employees who can be replaced at any time, and indeed will be at the mandatory retirement age of sixty-five. The committee system of collective management has long since displaced the autocratic style of John D. Rockefeller. Waiting in the wings are some five hundred managers whose professional progress and compensation are personally reviewed by the members of the management committee almost weekly. The one who survives this constant and intensive appraisal process will someday be chairman of Exxon. Most, although they will be well compensated, will be left wondering where they went wrong.

The goal of this management system is to obtain a consensus on how best to invest financial resources to earn profits and to avoid rash, costly mistakes. But even with this tightly controlled system of checks and balances, mistakes are sometimes made. For example, Exxon entered the nuclear business in 1969, and over the next decade its uranium-mining and fuel-fabrication activities lost $215 million. Complaining of counterproductive political decisions by the federal bureaucracy, the company decided to throw in the towel on almost its entire nuclear research and development program. Then, in 1973, Exxon and its partners paid $632 million to the federal government for the right to drill in the Gulf of Mexico off Florida. No oil was found and seven additional holes on which an additional $15 million was spent all proved unsuccessful.

Still, the record shows that Exxon's management has been more often right than wrong. Gross revenues for the first three months of 1980 were $28 billion, an increase of 102 percent over the first quarter of 1979, and cash flow came to some $200 million every week. Given the awesome economic responsibility generated by such figures, one hopes that the management committee of eight does not focus merely on its commitment to directors and stockholders. But no outsider really knows what the committee of eight is up to.

To know the men who run the show is not necessarily to love them. They are engaged in a tough business and are constantly judged by the bottom line. But a review of their diverse careers does refute the accusation that they are joined in a monolithic conspiracy, which in any case would be difficult to pull off, since their activities are subject to the scrutiny of the Justice and Energy departments and numerous federal and state regulatory agencies.

The same cannot be said of some others playing the oil game.

Chapter Six

Some Supporting Players in the Game

The Gnomes of Rotterdam

The 80,000-ton supertanker, registered in Liberia, owned by Greeks, and crewed by Filipinos, headed toward Rotterdam in the summer of 1979 with a cargo of Libyan crude oil. The cargo was especially valuable because Libya's very light, low-sulfur oil yields a high proportion of gasoline when refined. Sold through normal channels to a major oil company under a long-term contract, it would command the top official OPEC price of $23.50 a barrel.

This oil was not going to be sold at the official price, however. It was destined for the Rotterdam spot market. A full half-day before the tanker docked in Rotterdam's immense harbor with its backdrop of scores of oil storage tanks, the oil was sold to a West German refiner for $43.50 a barrel. Neither buyer nor seller talked to, much less saw, each other during the deal. It was all arranged by a trader operating over telex machine and three telephones in a small office without a nameplate on the door.

In another obscure office in Rotterdam, another trader received a telex message from a Scandinavian airline that desperately needed jet fuel at New York's JFK Airport. Moving with his customary speed, the trader arranged for a Caribbean refinery to ship gasoline to an oil company on the U.S. East Coast, which agreed to release a supply of its gasoline in Boston. The trader then moved to have this gasoline shipped to an oil company in Canada, which he knew had surplus jet fuel stored at JFK. The Canadians needed the gasoline more than the jet fuel, and exchanged it with the trader. The airline was assured of flying again and the trader, exhausted but assured of a healthy commission, went off to dine at Le Coq d'Or.

In yet another transaction, a trader received a message from a Soviet export agent in Finland. A Norwegian tanker was sailing from the Baltic Sea toward Rotterdam with a cargo of 26,000 metric tons of Soviet gas-oil, a heating fuel favored in Western Europe. Since the Soviet Union does not belong to OPEC, the gas-oil had no official ceiling price and was also not under long-term contract. After two days of frantic telephone and telex dealing while the tanker was still at sea, the trader sold the cargo to the highest bidder—a French distributor who sells gas-oil 150 miles down the Rhine—for about $9 million.

These transactions are typical of one of the most mysterious facets of the international oil business. It has become known as the Rotterdam spot market, although it involves brokers and traders in Monte Carlo, Paris, Oslo, Hamburg, Geneva, New York, London, Singapore, and elsewhere. They aren't registered, their transactions aren't reported, their price quotations are guesswork, and they don't really constitute a market because they don't meet at a bourse or on a trading floor. They do form a clubby, knowledgeable communications network with a turnover of billions of dollars a year that connects officials of oil-producing nations, executives of oil-producing and refining companies, distributors, tanker operators, brokers, and traders throughout the world. The worldwide spot market became associated with Rotterdam in the first place because of the city's port, refineries, oil-storage terminals, and nearness to Europe's pipeline system. Now the city's only reason for dominance is that prices—which affect every consumer of oil products—are set there.

The shadowy, lucrative spot market has a short history touched with irony. It originally came into being to even out imbalances between companies with too much or too little of an oil product, to provide buyers with cheap oil at a time of oversupply. In the summer of 1978, spot prices were below the long-term contract prices negotiated by the majors for bulk supplies because there was a glut of oil. Smaller companies could buy gasoline and sell it for less than it cost the majors to refine it. That was one good reason why the majors kept their gasoline prices in check.

The situation changed with the revolution in Iran and the consequent tightening of supplies. Buyers, especially the independents, were forced to scramble for oil that was not locked up under the majors' contracts, and they bid up the price. In the spring of 1979, OPEC's base price for a barrel of light Saudi Arabian crude, which was particularly suitable for refining into gasoline, was $14.55. On the open Rotterdam market, the same barrel sold for between $37 and $40.

The men who run this market are averse to publicity and do not grant interviews. In their business, a reputation for trustworthiness and an anonymity that will not jeopardize sensitive contacts within the market are everything.

Between two hundred and three hundred smaller traders are active in the spot market, and about fifteen major ones. Their companies have unfamiliar names, like Libra and Asmarine, Fearnley and Eger, Miske, Vitol Trading B.V., Lagon Trading B.V., Bulk Oil, and Coastal States Petroleum. Recently some of the major oil companies have begun to take an active interest in the spot market. Exxon with a company called Impco, British Petroleum with Anro, and Deutsche Shell with Petra Trading Company are all watching how things are going in Rotterdam.

This arcane trade of connecting buyers and sellers is not for the fainthearted. Prices are totally subject to the law of supply and demand and can change overnight, or even hourly. The only thing resembling New York Stock Exchange quotations are the five-day-a-week "assessments" carried by *Platt's Oilgram Price Report*, published by McGraw-Hill Inc. of New York City. This respected reporting service is obliged to obtain its information largely from the traders, brokers, and refiners themselves, who provide guesses at going prices.

Millions of dollars can be made in the spot market when prices rise dramatically, and millions can be lost when they fall. For example, a trader might find and close a deal for a cargo of crude oil still at sea for $25 a barrel, assuming that a buyer can be found at $30 a barrel, only to discover when the tanker docks two days later that slackening demand has caused the price to drop to $22 a barrel. However, in an era of steadily rising oil prices, the small group of traders in the Rotterdam spot market are operating in the world's most lucrative business.

The spot market affects Americans in a way most of them don't understand. For instance, gasoline was being sold for as much as $1.60 a gallon in the metropolitan New York area during the 1979 shortage. Those who bought scarce gasoline at those prices, and complained bitterly about the oil companies, probably had not even heard of the Rotterdam spot market. But that's where their gasoline came from, and its sale was legal.

Tankers and barges loaded with gasoline and other refined petroleum products came into New York harbor. The gasoline could have been refined anywhere outside the United States, from Canada to the Persian Gulf, and thus was not subject to the system of federal controls on domestic prices and distribution. The gasoline was then sold through word-of-mouth by an elusive network of brokers, some operating out of small offices near New York harbor close to Wall Street and others near Rockefeller Center. Eventually it wound up in the pumps of service stations whose allocations had been cut off by their regular domestic suppliers. Then the unregulated gasoline was sold for what the market would bear, after figuring in transportation fees, state and federal excise taxes, and the stations' profit margins (which *are* subject to federal regulation). It's doubtful that the spot market dealers had any sympathy for the Americans who were paying $1.60 a gallon for gasoline when pump prices in Europe were $2.00 or more a gallon.

Although only about 10 percent of the oil produced in the world is traded on the spot market, the spot market plays an important role in the international oil business because it acts as a barometer for the rest of the market. It has a definite effect on the prices of gasoline, home heating oil, and other petroleum products. After observing the high prices being paid at Rotterdam (almost double the official OPEC price), the OPEC producers (except Saudi Arabia) began adding surcharges of as much as $5 a barrel and demanding higher base prices.

The traders themselves are indifferent to consumers' welfare. They are simply interested in making a lot of money in a volatile business. Some critics of the major oil companies maintain that they encourage high spot market prices. According to their theory, only the majors have enough cash flow to afford these prices, and thus they are able to corner the market and pass the costs on to consumers. But this seems unlikely, for high prices on the spot market only encourage producing nations to place surcharges on long-term contracts with the majors and withhold oil committed under them in favor of quick spot profits.

The more moderate OPEC nations, led by Saudi Arabia, realize the danger in allowing oil prices to spiral indefinitely and ruin the economies of their principal customers. They have concluded a "gentlemen's agreement" to limit their spot transactions. Unfortunately, there is nothing to prevent countries who are not well-disposed toward Western democracies, like Iran or Libya, from breaking the agreement, abandoning long-term contracts, and diverting their oil to the highest bidder on a day-to-day open market when it is more lucrative to do so.

Most of the Western world's oil supplies are still handled by the Seven Sisters under long-term contracts with producing countries. The Sisters have generally been able to avoid the spot market, although they did buy a few loads of crude during the shortages of the spring of 1979 before pulling out when the price skyrocketed to $43 a barrel in early June. They chose to ration contracted supplies—however short they might be—rather than to compete for supplies at the Rotterdam prices. The independents, however, do not have the Seven Sisters' contracts so they have had no choice but to pay what the market will bear.

Anyone who decides to join the gnomes of Rotterdam, Geneva, and Hong Kong in playing this high-rolling game should be aware that the spot market can be a very tricky business. By the summer of 1980, it had softened because of conservation, a recession-induced fall in demand in the industrial countries, and high production in Saudi Arabia. Falling U.S. interest rates added to the downward pressure on European oil prices since crude oil, gasoline, and heating oil are sold for dollars on the world market. The majors found themselves, at least temporarily, stuck with bulging stocks of crude and high inventories of gasoline. Only a year after panic buying in the summer of 1979 had forced Rotterdam prices well above those contracted for by the oil-exporting countries and encouraged them to increase their prices further, an oil glut caused spot prices to drop below the $32-a-barrel ceiling charged for the OPEC benchmark crude, Saudi Arabia's light grade. Some of the African "sweet" crudes—produced mainly by Nigeria, Libya, and Algeria and traditionally priced higher than the Middle East crudes, which are generally heavier and higher in sulfur content—were bringing $37 a barrel.

Connoisseurs of irony will appreciate what happened next. In a reversal of roles, the small independent oil distributors who buy their supplies on the spot market were now able to do so at prices below those paid by the big oil companies, which were locked into long-term contracts with producer countries. The independents began underselling the majors, which had to respond by reducing their own prices for gasoline in Europe anywhere from 4 to 10 cents a gallon. For example, in Great Britain, Exxon and Shell gave their service station operators a "temporary" price cut of about 4 cents a gallon. In the United States, despite increases in the cost of controlled domestic oil, gasoline prices did not rise for a change. There is nothing worse for a business than an uncontrolled surplus of its product, and in the summer of 1980, the spot market—to which the insiders look—said that there was a surplus of oil and that it could be had for less than the majors were paying.

The situation was temporary, of course. In any commerce as vast and complex as oil, there are always temporary surpluses and shortages. With world oil production peaking for both geological and political reasons and demand continuing to increase, prices can only rise and shortages develop in the long run. The gnomes of Rotterdam will go on charging what the market will bear, thus encouraging the upward trend of contracted prices.

The Western oil-importing nations have become alarmed by this shadowy market, but thus far have not come up with any effective means to regulate it. "If you tried to regulate it in Rotterdam, it would just float away and resurface in Hong Kong or elsewhere," says John Swearingen of Indiana Standard. The traders can operate anywhere they can put willing buyers and sellers together.

The traders were back in action in force in the fall of 1980 when the outbreak of war between Iran and Iraq halted deliveries of between 3 million and 4 million barrels of oil daily. Despite comfortable oil reserves in most western countries, the threat of shortages caused heavy buying on the spot market at about 20 percent more by the end of October 1980 than prices paid immediately before the outbreak of the war.

There is only one way to control effectively the Rotterdam spot market and the economic dangers inherent in it. That is for the industrialized nations, led by the biggest oil consumer, the United States, to undertake serious conservation in the short run and a shift to alternative fuels in the long run. Only when oil demand falls significantly below supply will the hundreds of traders unplug their telephones and telex machines and fade into another line of work.

The Independent

Phil Niery also gambles. Not as a trader on the volatile Rotterdam spot market, but as a wildcatter here in the United States—an independent who searches for oil where none has been found before.

I arranged to meet him for lunch in the dining room of Houston's Petroleum Club. Houston is the oil capital of America, although recently Denver has assumed increasing importance as an energy center because it is close to the potentially large oil fields of the Overthrust Belt and the oil shale deposits of the west slope of the Rockies (three-quarters of future synthetic fuels development will probably take place in the Rocky Mountain and northern Great Plains states). But for now, Houston is America's oil headquarters and its Petroleum Club is the social inner sanctum of the industry.

As I walked through the club, I noted several oddities. Black chandeliers overhead. A large tapestry depicting the geographical strata of Texas. The deferential white-jacketed waiters who all seemed to be Hispanic. The absence of women diners. After exchanging greetings with Phil Niery, I asked him about the absence of women and he explained: "They're not allowed in here. But the wives of members can eat in the Discovery Room."

Glancing around the crowded room, I noted that most of the diners were dressed as conservatively as members of the Racquet Club in New York, although a few did wear leather boots and there were five Arabs in burnooses. Conversations were earnest and muted. I overheard a few strange phrases like "downdips," "heaving shale," "mud weight," "the Tuscaloosa Deep," and "the Deep Anadarco." The only celebrity present, as Phil Niery pointed out in a whisper, was a man who had once been married to the dancer Ann Miller. "She was born in Texas, you know," he said. "Saw her in *Sugar Babies* in New York. She can still tap."

Phil Niery lives in a modest Tudor mansion in the River Oaks section of Houston next to a residence that is a replica of Mount Vernon. He and his wife have flown to Leningrad to tour the Hermitage and to New York to attend the Metropolitan Opera, but where food is concerned, his tastes are unpretentious. He is known to favor souse, the head meat of the hog cooked with the gelatin of the feet and sprinkled with jalapeno sauce; I was relieved that day at the Petroleum Club when he recommended the chicken-fried steak, pecan balls, and a Lone Star beer.

Niery is a tall, wiry man in his middle fifties whose sandy hair is razor cut. He sported a blue suit of English cut and spectacles with thin gold rims. I had first met him at a meeting of the American Petroleum Institute sixteen years before and we had maintained the friendship in a desultory way. He had agreed to talk to me about his business provided I didn't print his real name ("Hell, you can quote my opinions, but you print my name in a book they're just gonna write in for money").

Niery is far removed from the boardrooms of Big Oil. He is of a little-known breed that is nevertheless important to the oil business and America's energy future. There are perhaps ten thousand independents who drill 70 to 80 percent of the new wells in the United States. They find nearly half of all new U.S. producing oil wells and almost three-quarters of all new natural gas wells.

It is a business known for its risks—"bummers" or "dusters" (dry holes) far outnumber the strikes and the chances are eight to two against finding oil in exploratory drilling. Nevertheless, the oil shortage and the end of price controls have made new drilling potentially highly profitable. Some categories of domestic oil brought nearly $40 a barrel in 1980, ten times more than in 1973. That dramatic rise in price made the game well worth playing. Julian Martin, vice president for research and administration for the Texas Independent Producers and Royalty Owners Association, thought that about fifty-four thousand new oil and gas wells would be drilled in the United States in 1980. (The all-time record is 58,160, set in 1956.) More investment funds are flowing to the independents through drilling funds similar to real estate syndicates. In 1979, members of the National Association of Securities Dealers raised more than $2 billion through 110 oil- and gas-drilling limited partnerships.

Independents like Phil Niery do their drilling on the mainland; offshore drilling, which is likely to yield any big finds that are left, is so costly that only the majors can afford to do it. Entering a limited partnership with even the most skilled independent is far from a guarantee of riches. Only two hundred of them gross more than $2.5 million a year. Niery is not in this league. And he has all but abandoned fantasies of emulating the "plays" of an H. L. Hunt or Sid Richardson. "The really big plays have already been made in the lower forty-eight. I'll accept that," he tells me. "But there's still oil and gas to be found . . . if the price is right.

And you need luck. In this game, it's better to be lucky than smart."

A native of Athens, Texas, where his father was a veterinarian, Niery graduated from Rice University in 1947 as a chemical engineer. He then joined a major oil company as a scout. Scouts obtain information on all leasing and drilling activities, by whatever means possible. The means can include spying on rival drilling operations through telescopes and field glasses and inducing competitors' roughnecks and toolpushers (drilling crew foremen) to talk while plying them with Jack Daniel's.

After five years as a scout, Niery was promoted to landman—a negotiator who buys drilling leases from private landowners. Six years later, he was promoted to supervisor at a Gulf Coast refinery. He then moved on to a marketing desk job at company headquarters in New York City. After twenty-one years, he was earning $29,000 a year and following the slow but steady career path rise of the typical large oil company executive. However, after having qualified for early retirement, he made a decision to go out on his own.

The year 1968 was not a promising one for a new independent. Imported oil was plentiful and much cheaper than domestic oil, whose production would peak in two years and then decline. Some independents were quitting the business, claiming that there was no longer enough money in it. But Niery moved his wife and five children from Westport, Connecticut, to Midland, a west Texas city of sixty thousand people whose tallest building is ten stories high and most of whose residents are in the oil business. There he found a geologist who thought he knew where oil still might be found and formed a partnership with him.

The next step was to find financing and secure leases. Here is where the independent plays his biggest role. He is the orchestrater of deals, the one who brings geological data, financial backing, and leases together and gets the show on the road. For financing, Niery put his own savings into the deal, got a loan from a Midland bank sympathetic to independents, and took in a third partner who owned two old drilling rigs. The geologist then charted the prospective terrain and Niery, acting as his own landman, went out and bought leases on it for a modest sum. The land turned out to have been drilled and abandoned by a major oil company, but the geologist was not daunted.

Before telling me what had happened, Niery recommended fried bananas for dessert and then wanted to discuss an important subject in Texas—football. This was an October Sunday and the undefeated University of Houston had triumphed over the previously unbeaten Arkansas Razorbacks the day before. "Just shows you how a school can come from nowhere with some backing," he said. "At this rate they'll have it all over Austin [the University of Texas] in a few years. Where I went, Rice [located in Houston], nobody seems to give a damn. But the city of Houston wanted a winner and now they've got one." Niery seemed to draw a parallel between the climb of the Houston Cougars from nowhere to the ranks of the country's top-ten teams and his own career.

"About that first play," he said, "that big oil company was right. It was a bummer. Of course, you do get a hundred percent tax writeoff for a dry hole. But for three straight years, we found nothing at all. My wife had to take a job as a waitress, my eldest son dropped out of SMU and went to work in a bank, and I wound up driving a truck on weekends. Then, in 1972, we bought a natural gas lease and drilled right down in the center of it. Nothing. An old toolpusher suggested whipstocking the field. That's drilling off at an angle. We did—and we hit. I've been making a good living with that and some other strikes ever since."

"Can you tell me how good?" I asked.

"Why not? It's not that spectacular and the IRS knows all about it. Last year, I grossed about nine hundred thousand and, after taxes, got to keep one hundred twenty thousand for myself and the family. Remember, most of that gross went into buying new leases and drilling new wells."

That struck me as a respectable sum for a small independent to risk in venture capital, but not spectacular when I recalled that for the first three months of 1979, Exxon claimed that its capital and exploration expenditures totaled $4.33 billion, with about 41 percent of it invested in the United States.

"What do you do with whatever you do find and produce?" I asked.

"It's bought by the majors, who refine and market it."

"But why do they need you and other independents in the first place? Why don't they just do it all themselves?"

"They do a lot of it themselves. But by dealing with us, they can have it both ways. With us doing the exploration work on the less promising leases, they don't take any risk and they don't have to keep big staffs on a permanent payroll. Anything we do find boosts their crude supplies. But let me tell you something you may not have realized. It's the majors, not the small independents, who own all those concessions overseas. We work strictly here at home. When all that cheap foreign oil was available, why should the majors have spent a lot of time and money looking for price-controlled domestic oil? No percentage in it. The majors still aren't hurting that much on foreign oil. Whenever OPEC raises its prices, so do the majors. Think about it."

"But isn't that all going to change with price decontrol?"

"It should. When the price is right, everybody will be scrambling for domestic oil."

"But I've been told there isn't that much left to be found."

"Maybe nothing really big, but there's still plenty. The point is we won't know until we start looking for it seriously. And nobody will look seriously unless the price is right. Never forget that this is a business."

"Isn't there a better way to do this?"

"You tell me how. Let the government do it? That would mean the end of the free enterprise system. You want another Amtrak?"

That response struck a familiar note. In researching this book, I discovered a paradox. The majority of Americans suspect oil men of fabricating shortages and reaping unfair profits. Oil men themselves, be they small independents, executives of Big Oil, or lobbyists for the American Petroleum Institute, see themselves as a misunderstood, beleaguered band, hamstrung by restrictive government regulations and the mountains of paperwork these require. If they have a common enemy, it is government. They would leave the solution of America's energy problems to the beneficent workings of the invisible hand of the free market.

"But eliminating all regulations and price controls has got to mean higher oil prices for other Americans," I said. "Phil, excuse me, but I think you're already doing all right. And you've still got percentage depletion."

"Big deal. So do most other extractive industries."

The percentage depletion allowance is still a sore point in the petroleum industry. It was reduced by the Tax Reform Act of 1969 from 27½ percent to 22 percent, and then eliminated by the Tax Reduction Act of 1975 for all but the smallest independent oil producers. In 1979, the percentage depletion allowance applied only to producers whose production of oil averaged below 1,200 barrels a day, like Phil Niery.

"You say I'm doing all right and maybe I am," he acknowledged. "But I'm not screwing anybody. I can't feel sorry that people are going to have to pay more for oil. They use much too much of it. And for years it was held below its true price by government controls, so they kept on lapping it up. All they'll be doing now is paying what oil is really worth. Remember that I damn near starved and nobody sent me any food parcels. I hustled and I took risks. Just last week myself and four other boys put together a deal to buy leases on four hundred thousand acres in the Appalachian Mountains. Now that's a gamble. I wouldn't have done any of this unless I thought I could make a good profit. Would anybody?"

"That's a fair question, Phil."

"My answer is nobody would. But I'm not in this just for the money. Sure, I'd like to become a millionaire, and maybe I will. But most of all I like being my own boss, being independent. There aren't many of us left in America. I even have a dream left. I'd like to find another Spindletop and get my name on a field like that. Put that in your book."

"How about becoming another Dad Joiner?"

"That's asking a lot. If I did, wouldn't that be good for the country? But you know he died broke."

"I read that someplace. You don't intend to die broke."

"No. Even if I come up with a string of bummers on these gambles I could always get a job someplace. Maybe as a lobbyist." He laughed.

"You don't think much of lobbyists?"

"I guess they're part of the game. But I don't really know much about what they do. They're mostly financed by the big boys, not people like me. Why don't you go up to Washington and find out for yourself? Then let me know. Washington is where the action is. How about another Lone Star?"

The Persuaders

The oil industry differs from other American businesses in one crucial respect. The major factor governing its operations is not sales to customers, who have little choice but to buy its products at whatever price (and to line up to do so during shortages), but the actions of state, local, and federal governments—especially the federal government. What Washington does, or does not do, about taxes, price controls, environmental restrictions, gasoline rationing, the leasing of federal lands for exploration, proposals to break up or nationalize the big oil companies, a national energy program, and hundreds of other legislative matters has a direct and vital impact on the industry and its very survival in its present form.

In the good old days, government treated the private oil industry kindly. It imposed quotas on then cheap foreign imports to protect the industry and the State Department and Central Intelligence Agency rendered vital support in foreign ventures. The companies enjoyed preferential tax legislation: percentage depletion, deductions for "intangible drilling expenses," foreign tax credits. Powerful congressional leaders like Sam Rayburn and Lyndon Johnson of Texas and Robert Kerr of Oklahoma could be counted on to stand up for the industry's interests.

An occasional incident did tarnish the industry's image, but never sufficiently to change its way of doing business. There was the Teapot Dome scandal, in which government officials transferred valuable publicly owned reserves to the oil companies. There was the antitrust suit brought by the Justice Department in 1941 against Exxon for exchanging vital information on patents and research with the German chemical combine of I. G. Farben after the Nazi invasion of Western Europe. Then there was that remark by Franklin Roosevelt's crusty secretary of the interior, which is echoed by many elected officials and citizens today.

During the war, Secretary of the Interior Harold Ickes gained the approval of President Roosevelt to form a government corporation called the Petroleum Reserves Corporation to buy from Standard Oil of California and Texaco their control of Saudi Arabian oil. Ickes's idea was that the government, rather than the private oil companies, should control this supply of oil for the benefit of the American people. His plan was quickly squelched by Socal and Texaco, who after the war accepted Exxon and Mobil as partners in the Arabian American Oil Company because they needed their sales outlets for the staggering glut of Saudi Arabian oil. Reflecting on his encounter with the oil giants, Ickes remarked: "An honest and scrupulous man in the oil business is so rare as to rank as a museum piece."

Later incidents did nothing to prove Ickes wrong. According to a prepared statement presented to the Senate Subcommittee on Multinational Corporations on July 16, 1975, by Archie Monroe, Exxon's controller, one Dr. Cazzaniga, the head of Exxon's Italian subsidiary (Esso Italiano), "had made unauthorized secret commitments and $11 million in unauthorized payments to SNAM, which was purchasing Exxon liquified natural gas (LNG) from Libya. SNAM is a subsidiary of ENI, an Italian State holding company."

Exxon's own internal investigation put the primary responsibility for the secret kickbacks on Dr. Cazzaniga, although at the subcommittee hearings Senator Church became curious as to how much Exxon's top management had known of his activities. The following dialogue took place:

SENATOR CHURCH: Well, did the head of the company, Mr. Jamieson, know?

MR. MONROE: At the time that Mr. Jamieson was contact for Esso on Europe he was aware we were making political contributions.

Mobil Oil Italiano, Shell, and British Petroleum made secret contributions to a wide spectrum of Italian parties and politicians. As it turned out, this practice was not confined to foreign politicians. The Watergate investigation turned up the illegal laundering of $100,000 donated to Richard Nixon's 1972 presidential campaign fund through a Gulf subsidiary in the Bahamas by Claude Wild, Gulf's vice president for government relations in Washington. After conducting its own investigation, the Securities and Exchange Commission charged Gulf with falsifying reports to hide a $10 million fund for political payments between 1960 and 1974. These payments were channeled through the subsidiary, Bahamas Exploration Ltd. Gulf agreed to refrain from such alleged violations in the future in a consent decree.

During the hearings of the Senate Subcommittee on Multinational Corporations, B. R. Dorsey, chairman of the board and chief executive officer of Gulf, denied having had any knowledge of secret political contributions before Claude Wild told him about them during the Watergate investigation in July 1973. Senator Symington of Missouri found this hard to believe.

SENATOR SYMINGTON: How was it expressed on the balance sheet you would sign when you put out your earnings statement to your stockholders? . . . Did you put it under the heading of miscellaneous?

MR. DORSEY: Miscellaneous expense.

SENATOR SYMINGTON: And there were no questions about what this miscellaneous expense was for?

MR. DORSEY: Senator, this was a relatively small amount of money. During this period of time I think the company did some $60 or $70 billion worth of business in that 15-year period, and $10 million is not really a very large amount of money, it does not stand out.

Everything is relative, of course. To a chairman of the board of a big oil company, perhaps $10 million is not a large amount of money. An outside review committee hired by Gulf's board of directors concluded that Dorsey was "not sufficiently alert and should have known that Wild was involved in making political contributions from an unknown source. . . . If Dorsey did not know of the nature and extent of Wild's unlawful activities, he perhaps chose to close his eyes to what was going on. Had he been more alert to the problem, he was in a ready position to inquire about it and put an end to it."

The board asked for and received B. R. Dorsey's resignation. A board member, Sister Jane Scully, president of Carlow College, commented: "I felt enormously sympathetic to Bob Dorsey. He is a good and decent man, a fine person. But it was imperative the corporation restore its own sense of rectitude."

These and other incidents that came to light over the years gave the industry a rather bad image. As early as 1971, Frank Ikard, president of the American Petroleum Institute, acknowledged: "Quite frankly, the oil industry has developed the reputation over the years of being a robber, cheating and despoiling the environment." Nevertheless, as long as the conduct of this powerful and influential industry was not too publicly outrageous and it used its expertise to bring voters plentiful supplies of cheap gasoline and other oil products, the hand of government regulation remained relatively light.

The watershed in Big Oil's relationship with the public and the government was the Arab embargo of 1973–1974. With the subsequent quadrupling of oil prices, angry lines at the gas pumps, and rising company profits, there were anguished cries from the lawmakers' constituents, who perceived the industry was using its power to rip them off. The turnaround from cheap, plentiful energy was too swift for the society to absorb, and the government's reaction to the oil shock was to protect consumers from price increases by controls on prices.

An aroused Congress investigated oil company profits and shortages and proposed legislation that the industry viewed as "punitive." New regulations were imposed by federal bureaucrats. A program of "entitlements" was designed to equalize the costs of oil and gasoline around the country. Individual refineries, which had previously relied on either domestic or imported oil, were allowed to process price-controlled domestic oil only if they agreed to purchase some high-priced foreign oil. (Strangely, this system actually promoted imports at a time when the government's stated policy was "energy independence.") The oil companies were able to defeat a windfall profits tax proposed by President Ford in 1975, but by the end of the decade, about the only influential allies they had left on Capitol Hill were Senator Long of Louisiana and Senator Lloyd Bentsen of Texas, lawmakers from the two biggest producer states. Long, who has been in the Senate since 1948 and is often called its most powerful member, was chairman of the tax-writing Finance Committee. Both he and Bentsen are Democrats, but the Democratic president, Jimmy Carter, was no friend of the industry. Perhaps aware of the old populist appeal in attacking Big Oil, the president in 1979 pronounced oil profits "excessive," "undeserved," and "enormous."

All these attacks greatly angered the oil executives, who continued to claim that their profits were not large enough to do the job of assuring oil supplies for the nation. Their front line of defence in Washington was the oil lobby, whose members are as little known to the public as are the traders on the Rotterdam spot market. The oil lobby is the industry's principal instrument for getting its message across to the legislators who are responsible for the passage of laws on petroleum, laws that directly affect all American consumers as well as the balance sheets of the companies that sell the oil. The lobby, like the industry itself, is by no means monolithic. Nor is it illegal; all major industries have lobbyists in Washington looking out for their special interests. There are even consumer lobbying groups, although they obviously don't command the financial backing that an Exxon or Mobil can supply.

Oil and gas interests have long been a rich source of political contributions. Between 1973 and 1977, they contributed nearly half a million dollars to members of the Senate Finance Committee, and in 1978, they made $1.3 million available to thirty-four senators who were up for reelection. While there is no evidence that such spending buys any votes, it certainly ensures that a lobbyist will get a hearing in the Senate.

The lobbying corps consists of more that six hundred petroleum industry employees, as well as platoons of lawyers, public relations and advertising specialists, and other consultants on retainer. Their activities cost the industry somewhere between $10 million and $75 million a year. An exact figure is impossible to determine because of the way disclosure laws are written. Mobil, for example, spent $3.3 million on TV and newspaper "advocacy advertising" in 1978, none of which had to be reported as a lobbying expense under the law.

The elite of the lobbying corps represent the giants—Exxon, Mobil, Texaco, Shell, Standard Oil of California, Gulf, Standard Oil of Indiana, and Atlantic Richfield. David Gross of Shell, a lawyer, has a staff of twenty-three; Paul Petrus of Mobil, a chemist by training, has a staff of sixteen; and William Tell of Texaco, a lawyer, has a staff of ten. In 1980, the dean of this elite was Donald Smiley, a forty-nine-year-old Exxon vice president in charge of its Washington office with a staff of six. Smiley reported to William Slick, Jr., senior vice president for government and public affairs in Houston. Together they form what is sometimes known as the "Slick and Smiley Show."

In the good old days, a lobbyist could sit down with a committee chairman in a friendly club and, after a few bourbons, make a deal. The lobbyist might suggest a campaign contribution or the building of a job-creating petrochemical complex in the chairman's state or district in exchange for the passage of a law favorable to the lobbyist's company. Today the techniques are more sophisticated. Campaign contributions have been restricted by recent federal legislation, although they can still be made in reduced amounts through business political action committees (PACs) and can be useful in opening doors. For example, Amoco reported contributions of $249,200 to various candidates in the 1978 election through its PAC.

Instead of simply cultivating senior committee chairmen, Messrs. Slick and Smiley and their fellow lobbyists zero in on all 535 members of the House and Senate, as well as on members of the administration. Either through personal contact or by circulating well-researched reports and position papers, they try to explain their industry's and their companies' positions on a variety of public issues. Their aim is to provide detailed, reliable information on complex issues to busy legislators who often don't have the staffs to assemble such information themselves. Especially useful information relates to how some proposed energy legislation would affect the legislator's home district.

Like the Rotterdam spot market traders, the lobbyists' reputation for trustworthiness and discretion is all important. While they are obviously working to further the interests of the oil business, the facts they present must be accurate. If, when faced with the facts, a legislator still protests that his constituents are complaining about high oil company profits, rising prices, and shortages, the lobbyist diplomatically reminds him that consumers are contradictory in their demands for action. They want others—like the legislator and the oil companies—to do something about energy, but seem unwilling to do much about it themselves. They don't want gasoline rationing, won't practice conservation, don't like car pools, and refuse to accept the reality that an effective solution to the energy crisis means higher prices or lower consumption or both. And then there is an obvious fact of life that rarely has to be stated to a politician: that the oil business has hundreds of thousands of employees and stockholders and that they vote, too.

Senior executives of the majors also engage in lobbying by visiting Capitol Hill to explain their views to congressmen, calling on officials of the Department of Energy, and writing letters to the president and his aides. Other organizations with considerable clout, like the National Association of Manufacturers, the Business Roundtable, and the Chamber of Commerce, agree with the oil industry on many issues, and they lobby accordingly.

More visible to the public eye than the lobbyists of the major oil companies is the American Petroleum Institute, the trade association that represents the industry's views before Congress and with the Department of Energy. The API has 350 corporate and 7,500 individual members, 500 employees, and an annual budget of $30 million. It publishes an array of production figures that, astonishingly, were until recently the only industry production figures the Department of Energy had available upon which to base its recommendations. Originally, the API merely collected data and acted as a clearing house, but in 1970, it went into lobbying and analyzing public issues and policy.

Since January 1979, the president of the API has been Charles DiBona, who was born in Quincy, Massachusetts, in 1932, graduated from the U.S. Naval Academy in 1956, and was a Rhodes scholar at Oxford. DiBona has served on submarines and destroyers, was a whiz kid in systems analysis at the Pentagon in the early 1960s, resigned from the Navy to become president of the Center for Naval Analysis, and then led a staff examining energy issues at the White House in the early 1970s.

"Congress is concerned about how their constituents react to problems," DiBona says. "We're trying to deal with that extended constituency. We hope to convince people that some of the things we're trying to do are in their interest." He realizes that this is no easy task: "The problem of industry credibility has been so severe that many people believe that our mistakes are intentional and misleading. So we spend a lot of time trying to check what we say."

It's not easy to achieve a consensus among the diverse groups that belong to the API. The independents and the integrated companies, the domestic companies and the major international companies, often have different objectives and opinions on how to achieve them. But the API is where oil men at least try to coordinate their lobbying. "We try to find issues that we can analyze, then find a consensus in the industry, and then argue for it," DiBona says.

The actual lobbying is done by Charles Sandler, vice president of the Office of Government Affairs, who has a staff of six and an annual budget of about $1 million. Sandler claims that there is nothing mysterious about his work, which he describes as "a hard, nuts-and-bolts job." He and his staff cultivate congressional aides and let the lobbyists of the majors cultivate the congressmen themselves.

Ranking below the elite troops of the majors and the API in the hierarchy of lobbyists is a host of persuaders for the independents, jobbers and refiners, explorers, pipeline operators, and truckers. Dozens of groups represent specific interests, like the American Gas Association, the National Oil Jobbers Council, the Domestic Petroleum Council, and the Council of Active Independent Oil and Gas Producers. The API maintains petroleum councils in most state capitals and works with such organizations as the Mid-Continent Oil and Gas Association, the Western Oil and Gas Association, and the Rocky Mountain Oil and Gas Association. The Independent Petroleum Association of America, whose executive vice president is the Oklahoma-born former newsman Lloyd Unsell, has fifty-one hundred members in thirty-three states and a Washington staff of thirty-two.

When their common interests are affected, all these organizations can put a lot of pressure on the congressmen who represent their home state and district. Hundreds of their members send telegrams, make phone calls, or descend upon Washington in person in their Lear jets from Texas and Louisiana and Oklahoma.

In addition to this formidable array of lobbyists (although, according to Unsell "the terror and the power of the oil lobby is the biggest myth in the country"), there are others who are not employees of the companies or associations but who represent them for a fee. They are usually lawyers or former congressmen with good contacts on the Hill. One of the leaders in this lucrative field is the law firm of Smathers, Symington, and Herlong, which represents the Pennzoil Company of Houston. Symington is a former representative from Missouri; Smathers is a former senator from Florida. Another is the law firm of Thomas Hale Boggs, Jr., which has among its clients Marathon Oil Company, the Association of Oil Pipelines, and the Independent Tanker Owners Association. Boggs is the son of the late T. Hale Boggs, Democrat of Louisiana and House majority leader.

How effective is this horde of lobbyists whose activities are paid for by all Americans who buy oil products? It is difficult to judge, for they decline to talk about their triumphs or defeats. In their best of all possible worlds, there would be no government regulation or taxation of the oil business at all. But in the real world, that is not meant to be, so they strive to make the taxes less onerous and the regulations less restrictive. It's safe to say that they played a role in persuading the Carter administration and Congress to phase out price controls.

One story illustrates how the lobbyists go about their work. It involves President Carter's national energy program as submitted to Congress in 1978. An important feature of the bill was a tax on industrial users of oil that was meant to force them to switch to more abundant coal.

The Lone Star Steel Company saw economic hardship in this proposal and hired Charls Walker to fight it. Walker was deputy secretary of the treasury in the Nixon administration and now has his own Washington consulting firm. He convinced the managements of Union Carbide, Eastman Kodak, and Du Pont that the proposed tax would be bad for business. With this powerful backing, he persuaded Representative James Martin, Republican of North Carolina, of the same thing.

The textile industry was also against the tax. So Walker went to a representative whose constituency includes the textile industry, Kenneth Holland, Democrat of South Carolina. Representatives Martin and Holland then introduced amendments that weakened the tax proposal before it went to the Senate.

Walker wasn't finished. He next called on members of the Senate Finance Committee, headed by the powerful Russell Bilieu Long. Many committee members didn't think much of the tax themselves. To bolster their misgivings, Walker got another of his clients, the Business Roundtable, which is an influential group of corporate leaders, to join the lobby. Eventually, the Senate Finance Committee killed the proposed tax altogether. President Carter may have had this particular lobbying effort in mind when he criticized "the enormous power of a well-organized special interest."

That power, however, was not strong enough to prevent passage of the piece of legislation against which the lobby put up its fiercest battle—the windfall tax on oil profits that was the financial keystone of President Carter's energy-sufficiency program. Simply stated, this complicated bill imposes a tax through 1990 on the huge profits the oil companies will realize from the decontrol of domestic oil prices.

Though they failed to stop the bill altogether, the oil lobby did manage to dilute it in the Senate so that the expected revenue to the treasury will be only a little more than half the amount originally asked by the administration and about two-thirds that amount provided for in a measure approved by the Ways and Means Committee of the House in June 1979. The successful effort to reduce the tax was orchestrated by the redoubtable Senator Long in what was described by the *New York Times* as a "bravura performance." The final measure also allowed the oil companies to deduct windfall payments against their regular corporate taxes.

Senator Long did not accomplish this feat solely because of his ingenuity and persistence. When we speak of a national energy policy that works to the benefit of the American people, we must take into account the power of a new group that has jokingly been termed OOPS—the Organization of Oil Producing States. There are thirty-one of these states, ranging in production capacity from Texas, with almost 4 million barrels of crude a day, to Virginia, with 12 barrels a day. Many of these states are populous and growing more so every day as the great migration of Americans to the Sunbelt continues. Voters in these states know the energy industry has boosted their economies and do not necessarily see their interests as coincidental with those of Americans in other parts of the country. The legislators representing OOPS have a potent voice in determining national energy policy.

Relating to the Public

In addition to employing hundreds of lobbyists to roam Capitol Hill buttonholing legislators, the oil business tries to exert influence through public relations. The industry once scorned p.r. as mere flackery, in which it had no need to engage. Now all the big companies (except the loner, Amerada Hess) maintain sizable public relations departments of as many as forty people headed by vice presidents and senior vice presidents. They deal with TV and newspaper reporters; churn out press releases, position papers, and fact sheets; publish magazines for employees, stockholders, and retailers; write speeches for senior executives, orchestrate tours and seminars for academics and community leaders; make informative films; contribute money to worthy causes like university research; and underwrite cultural projects like radio broadcasts of the Metropolitan Opera and BBC dramas on public service television.

All this activity is designed to improve the industry's battered image and to demonstrate to a suspicious public that the oil companies are good guys. In addition to their own public relations departments—now often designated by the more elevated title "public affairs"—the majors employ outside public relations agencies. Hill and Knowlton, the largest, has Texaco and Atlantic Richfield as clients; Burson-Marsteller, the second largest, has Indiana Standard; and Earl Newsom and Company has Exxon.

The services these outside agencies provide are a startling departure from the standard chores of helping a client with annual reports, trade shows, and product publicity. Many agencies have their own mock television studios where an executive can prepare for an upcoming interview by sitting under lights and fielding questions thrown at him by an angry panel (Why are oil company profits so big? Why did you hold back supplies to wait for higher prices? Why don't you do something about the Arabs?). The executive's performance is videotaped, then analyzed and refined. At Burson-Marsteller's Chicago office, managers of Indiana Standard have been run through something called Crisisport, a mythical world where actors confront them with tough and unexpected situations.

Some oil companies are making a determined effort to educate and motivate their employees in the political process. Exxon, for example, holds twenty two-day seminars a year for about fifty employees each in which they watch videotapes of speeches by politicians and a simulated campaign game called "See How They Run," and learn how mileage standards were built into the Clean Air Act. Atlantic Richfield employees, during an eight-hour political education seminar, role-play at running political campaigns in the fictitious state of "Erewhon." The goal of these programs is to motivate employees to work at the grass roots for the companies' interests. Although they are voluntary, and no one is forced to spout the company line, one wonders how far rising young managers would continue to rise if they declined to participate with enthusiasm.

Another favorite public relations ploy is the faculty forum or seminar in which oil companies invite college professors to their headquarters for frank discussions with their executives about the energy situation. Sometimes, however, this effort to educate the educators backfires. The following article by George H. Wolfe, assistant dean of the University of Alabama's College of Arts and Sciences and associate professor of English, printed in the *New York Times* of December 27, 1979, throws some light on the task facing oil men in their public relations struggle for credibility.

> Recently in San Francisco I realized why some of us not in the oil business have such difficulty believing fully in the oil crisis.
>
> Representing my university at a meeting whose host was Standard Oil of California (Chevron), I was struck by the disturbing similarity between this so-called faculty forum and one I'd attended earlier in Houston sponsored by Exxon. Both were characterized by seeming good cheer, both paired groups of youngish academic humanists together with mid-range fast-risers in the oil company, both appeared designed to facilitate honest discussions of the energy situation, but both proved to be elaborate pseudo-events, and therefore wastes of time. Perhaps worse.
>
> That companies the size of Exxon and Chevron, with budgets and clout greater than half the member nations of the United Nations, feel compelled to sponsor these meetings at all puzzled me. What do they hope to gain? Academics are notoriously easy to overwhelm with mammon. The average salary of a 35-year-

old associate professor of history hovers around $18,500. Theoretically, we can be easily bought. Academics know this and the better ones tend to develop, amid sponsored opulence, rather a keen skepticism of their host's generosity. The first evening, the oil men encouraged us to be frank. They promise no-holds-barred, and, despite our caution, we are secretly flattered to have been invited to consider some sticky social problems facing America.

Humanities scholars tend to be advocates for ideals and abstractions, not purveyors of market solutions. They are not salesmen. They are not better than salesmen, but they are different. The oil men *are* salesmen, drummers whose product is at the moment in extraordinary demand and whose sales policies are undergoing microscopic scrutiny by an enraged government and its people. Hence, their keen interest in their own public image and the ideology of mega-profits.

As the forum gets under way, these differences immediately assert themselves. It is clear that the host's representatives (all men) are on home turf. This is their subject, and, while they are routinely affable and polite, when talk veers toward controversy they close ranks with precision. Assume that a professor blunders through a question about decontrol, or divestiture, or the old-oil/new-oil scam, or unconscionable profits, or the Mexican Connection, or the paying of baksheesh to foreign governments, or illegal political contributions, or whatever; the possibilities are myriad. In the first place, even if the professor has prepared himself by slogging through the latest accusations of Big Oil's malfeasance, it is likely that his grasp of a specialized terminology will fail him at a crucial moment, and he will be skating on exquisitely thin ice.

It is not unusual, certainly, for a company line to emerge from a company man, but this is something bigger, an industry line being pressed upon a group of apparently powerless academics who have great difficulty distinguishing between propaganda and fact, between lunacy and legitimate solutions. One realizes that he is being lobbied. The ostensible purpose of open, frank discussions of The Energy Situation now seems facile and one's own naïveté embarrassing. What did you think you were out here for?

But the question remains: why bother? A group of almost

pathetically uninformed professors can never penetrate the intricacies of an industrial labyrinth unexampled for complexity in the modern world. And that is, of course, the point: We are there to serve as razor strops. The oil companies assume that as a group we are educated, articulate and only occasionally combative. Although we may abhor the greed of Arab sheiks and suspect the oil giants of duplicity on a grand scale, we are likely to be fairly civilized in our disagreement. What better group to pit against company spokesmen whose aims appear two-fold: to present their version of the history and current state of the energy crisis, and to try and find out what The People think about them and their multinational dealings—without revealing anything significant about their operations.

In the end, the faculty forums failed for the professors because, by our lights, the oil men cheated on the deal. Never mind what we should have expected; what we got was a sham. In exchange for a couple of days in the city, we served to update a small, relatively trivial segment of the company's defense repertoire.

All things considered, I'd rather have been in Philadelphia.

Chevron probably also wished Dean Wolfe had been in Philadelphia.

Another segment of the industry's defense repertoire is advertising, and the money spent on it is not trivial. Gone are the days when the majors ran hard-sell campaigns urging us to put a tiger in our tank; they don't have to hustle gasoline anymore. Now the ads are what is known as "institutional." In 1979, Exxon spent $14 million on advertising that tried to sell the company's opinions on a wide variety of energy-related issues. On TV, Bob Hope assures us that Texaco is working to keep our trust. A study by the General Accounting Office calculated that Exxon, Texaco, Gulf, Mobil, Indiana Standard, and Shell spent $425 million on advertising from 1970 to 1972. Most of this was tax deductible, since the companies maintained that the ads were of a "corporate image" and "informational" rather than political nature.

Mobil is easily the industry leader in public relations or "public affairs." In 1978, the company had a public affairs budget of $21 million, about 25 percent of which went for print ads that appeared in magazines and opposite the editorial pages of leading newspapers like the *Washington Post* and the *New York Times*. They vigorously asserted Mobil's position on almost everything affecting energy and debunked the industry's critics with well-researched facts.

This massive effort is headed by Herbert Schmertz, vice president for public affairs, who has a staff of nearly one hundred people. Schmertz, of course, does not run the show at Mobil (Chairman Warner and President Tavoulareas do that), but he does run the biggest public affairs program in the history of the industry with the backing of top management.

Schmertz does not come from the Southwest, wear cowboy hats, or say "Howdy." He is far from being the archetypical oil man with political views somewhat to the right of H. L. Hunt (which is probably why he was hired). He was born in Yonkers, New York, where his father was a jeweler, went to New Rochelle High School and Union College in Schenectady, New York, and graduated from Columbia Law School in 1955. After two years in the Army, he became a labor arbitrator and monitor of union elections.

In 1960, Schmertz helped with voter registration in the campaign of John F. Kennedy and was then made general counsel of the Federal Mediation and Conciliation Service. In 1964, he returned to private practice and two years later became manager of labor relations for Mobil. In 1968, he took a leave of absence to work as an advance man in the presidential campaign of Robert Kennedy. The next year he was made vice president of public affairs, and after a stint in charge of Mobil's fleet of tankers, returned to the public affairs arena in 1973 when the flak from the Arab embargo hit Big Oil.

Schmertz designed a clever but simple strategy to counter the criticism. "Our motivation is not to have people love us," he said. "It's more complex. We want to stimulate and participate in dialogue and want people to take us seriously. . . . We are doing an enormous amount and we're doing exactly what we're trying to do. We may fail, but we know what we want to do."

Not that it was going to be easy. "Through the 1960s and right to this day," Schmertz declared in a speech before the American Iron and Steel Institute in September 1978, "a strong undercurrent of anti-business sentiment has made it difficult for corporations to get a fair hearing in the forum of national debate. . . . The academic community has trained nearly a generation of Americans in a catechism which holds that 'big is bad.' " He added: "Sad to say, the free American press—print and electronic—has been caught up in this anti-business maelstrom."

To get a fair hearing in the maelstrom, Mobil's strategy was to lay down a barrage of advocacy ads in respected publications along with letters to the editor advancing its opinions and contesting any contrary to its interest or that of the oil business in general. Through its paid advertising, Mobil's copywriters became, in effect, nameless columnists for the *New York Times*. The company placed a newspaper ad in every congressional district when the decontrol issue was before Congress. These ads are aimed at legislators and the more literate voter who might be swayed to a pro-oil point of view.

For those unwilling or unable to plow through the op-ed ads, there are the brief, easy-to-read comments called "Observations" that appear in Sunday supplements, often enlivened by cartoons (Schmertz commissioned Charles Addams to draw a few of them). And, for the culture-hound, there are large grants to public service television to support British-produced serial dramas on *Masterpiece Theatre* such as *Upstairs, Downstairs* and *I, Claudius*. Mobil has unquestionably upgraded the quality of public TV; *Masterpiece Theatre* has won sixteen Emmys. While this may seem a long way from what Dad Joiner and his roughnecks were up to in the old east Texas oil fields, Mobil does have a practical motive: It hopes to convince influential people that it is a sensitive outfit and thus make them more receptive to the company's messages.

As a patron of the arts, Mobil has won some friends by, among other things, supporting free admission and late hours at the Whitney and Guggenheim museums in New York City and the operation of the Summergarden at the Museum of Modern Art.

Mobil even sponsors books about regions where it has petroleum interests, such as *The Genius of Arab Civilization,* published by New York University Press.

Herb Schmertz has encountered some problems and stirred up controversy in his role as a twentieth-century Medici. The commercial television networks have spurned his largess, contending that Mobil's "idea" ads would draw demands for free equal time and that controversial issues should be dealt with in their news and public affairs programs. Schmertz complained this was censorship and offered to buy equal time for his adversaries (the networks declined). He even considered forming his own network. Meanwhile, in early 1979, he assembled a network of forty-nine independents and CBS, NBC, and ABC affiliates to broadcast the *Edward the King* television series (about Edward VII) that Mobil had bought from England and the networks declined to air. Accompanied by subdued Mobil commercials, *Edward the King* won the ratings battle against the fare broadcast by the networks in the same Wednesday 8 P.M. time slot.

In early 1980, Mobil syndicated a six-part British-made series called *Edward and Mrs. Simpson,* about the romance that led to King Edward VIII's abdication. Inserted at plot intervals were commercials in the style of "fables." Three stations owned by the Washington Post–Newsweek broadcast group banned the ads, and two consumer groups, Energy Action and the Citizen/Labor Energy Coalition, mailed a letter to thirty stations that did broadcast *Edward and Mrs. Simpson* demanding "equal time" for what they saw as the "overtly political" Mobil ads.

Mobil's chairman, Rawleigh Warner, Jr., was undismayed by the furor. In discussing the successful reception of the series about Edward VII, he told a columnist from *New York* magazine:

> I think we win friends. I think there's an intelligent audience out there that would go so far as to say, "You know, I don't like those characters, but if they give us this, maybe I ought to read what they have to say, because maybe they do have something to say."

I think it's perfectly appropriate to attract the attention of the audiences we're trying to reach, because those that are against us are certainly utilizing every weapon at their command to reach the same audience.

The sophistry of this view is that those who want to take an objective look at the operations of the oil companies, or even to criticize them, do not command the weapons that Mobil's financial resources provide. Nevertheless, illuminating confrontations do occur. One of them, which became rather testy at times, illustrates the hazards lying in wait for even the best planned and financed public relations effort. Conducted in the letters to the editor column of the *New York Times,* it provided an unusual insight into the real financial workings of the money game and gave those who followed it a chance to make up their minds as to where the truth lay.

The confrontation began on May 31, 1979, when Mobil ran an ad entitled "Chilling the Debate," which sternly took to task several groups, among them the Citizen/Labor Energy Coalition, and elected officials, among them Senator Thomas Eagleton of Missouri, who had had the temerity to criticize Mobil and the industry. Senator Eagleton did not stand still for that and replied in a letter of June 12, 1979:

In a recent advertisement in The Times, the Mobil Oil Corporation attacked me for my statements against domestic crude oil price decontrol. If I had Mobil's millions for advertising, I would respond in kind. Since I do not, a letter to the editor must suffice.

Specifically, Mobil suggested that I mislead the public when I point out that American oil companies now spend about $1.50 to pump a barrel of domestic crude and then sell the same barrel of oil for more than $8.

If Mobil wants to excoriate someone for circulating these figures, it needn't search as far as the U.S. Senate. It can find a suitable culprit in its own accounting department, for the figures are Mobil's own.

In its 1979 Form 10-K Annual Report to the Securities and

> Exchange Commission, Mobil lists its domestic crude oil "average production cost, as defined by the S.E.C." as $1.52 per barrel. The corporation lists its own "average price/transfer" value as $8.33 per barrel, for a per-barrel markup of 448 percent. In a footnote, Mobil suggests that the S.E.C.'s definition reflects only half of actual costs. Even if one accepts this argument (which Mobil alone among all the oil companies makes), the corporation's markup on a barrel of crude is more than 174 percent.
>
> To some cynical souls, all of this might seem to be just another case of the kind of rubber bookkeeping which has been prevalent in the oil industry. For example, in its critical advertisement Mobil also decries claims that "Mobil's effective tax rate on U.S. income was 21.2 percent." The ad states that "Mobil's *actual* effective U.S. tax rate on U.S. book income [was] 41.9 percent." Incredibly, in the very next paragraph Mobil goes on to note that on a barrel of domestic oil selling for $8.33 "about $1.35 is paid in taxes." For those who left their slide rules at home, $1.35 is about 16 percent of $8.33. Once again, Mobil Oil turns out to be its own most damning accuser.
>
> I continue to see in these figures little justification for giving the oil companies a bonanza of unearned profits as incentive to pump old crude.

Obviously, Mobil's Washington lobbying corps had not made much of an impact on the incredulous Senator Eagleton. But Mobil was not about to let the senator get away with his argument, and on June 22, a rebuttal appeared from James Q. Riordan, a senior vice president:

> . . . Senator Thomas Eagleton reveals his continuing confusion about oil company profits, and he misstates Mobil's position on crude oil decontrol.
>
> The Senator's confusion stems from certain information about crude oil costs and prices that Mobil (like other oil companies) was required by the Securities and Exchange Commission to publish in the Form 10-K. He cites figures of $8.33 for the average sales price of a barrel of U.S. crude oil and $1.52 a barrel for "average production costs," published in

accordance with a narrow definition mandated by the S.E.C., and concludes that Mobil enjoyed a 448 percent mark-up, leading to a bonanza of unearned profits.

When the Senator first used S.E.C. figures to accuse us of abnormally high profits, we were not surprised by his confusion. We had told the S.E.C. that its new reporting requirements might confuse people. We had tried to avoid the confusion by putting a footnote in our 10-K report explaining that the difference between the two S.E.C.-mandated figures "does not represent profit to the company" since, in accordance with the S.E.C. definition, the "average production cost" excludes expenditures associated with finding and developing oil, as well as income taxes.

When Senator Eagleton and others in Washington persisted in misinterpreting these figures, we announced that Mobil's average profit per barrel (for crude oil and natural gas liquids plus natural gas on an oil-equivalent basis) was approximately $1.50 in 1978. For Senator Eagleton to continue mentioning a 448 percent "mark-up" after all this is disheartening.

Not content to misread and misuse S.E.C. data, Senator Eagleton's letter indicates that he has also gotten himself confused about Mobil's effective income tax rate. It will come as no surprise to Times readers that the U.S. imposed income tax on corporations at a statutory rate of 48 percent until this year (now 46 percent). The law permits taxpayers to have the benefit of such items as the investment tax credit, which reduces income tax liability for most corporations to less than the statutory rate. In 1978, Mobil's provision for U.S. income tax ($385 million), when compared with Mobil's U.S. book income ($577 million), gives an effective rate of 40 percent, as we pointed out in the Op-Ed ad to which the Senator was responding.

Senator Eagleton suggests that Mobil's effective rate is 16 percent, and to prove his point he introduces a rather bizarre approach to income tax calculations: relating income tax to selling price instead of income. Specifically, he takes the figure of $8.33—the average sales price of U.S. crude oil—and divides that into $1.35 (the estimated income tax provision per barrel).

It is true that $1.35 is indeed 16.2 percent of $8.33, but the calculation is meaningless. The fact is the U.S. does not impose an income tax on sales revenues. It imposes an income tax on income. From all of this, I can only conclude that it is hard to do your sums when you are hunting scapegoats.

Finally, when Senator Eagleton says he sees in his interpretation of our figures "little justification for giving the oil companies a bonanza of unearned profits as incentive to pump crude oil," he once again takes aim at the wrong target.

He conveniently ignores an essential point in the energy proposal introduced at our annual meeting of shareholders in his home state on May 3. Mobil's president, William P. Tavoulareas, proposed there that the oil industry should *forgo* for now any price increase beyond inflation on oil from wells *currently in production,* providing the industry is allowed the world market price on oil *it discovers in the future*. This would eliminate any possible "windfall" on the "old crude" mentioned by Senator Eagleton. "Certainly," as Mr. Tavoulareas pointed out, "there can be no windfall profit on oil that has not yet been found."

We had hoped with this compromise position to help the energy debate move from political rhetoric to rational discussion. We made our proposal because we knew that the mere mention of so-called windfall profits made some people so mad they couldn't count straight.

Senior Vice President Riordan's rational and well-documented letter about taxes should have enlightened the confused Senator Eagleton and ended the debate. But it didn't. On June 28, Heather Booth, executive director of the Citizen/Labor Energy Coalition, joined the fray:

Mobil claims that our assertion that they pay very low income taxes is wrong, that they have tried in vain to determine where we got our figures, and that their *real* Federal income tax rate is 41.9 percent. They tell us to look at their annual report for the documentation. We did.

If the Mobil public relations department would only look at the company's own annual report they would discover that Mobil's net U.S. income (after state and local taxes and overhead) was $1 billion in 1977. Their current Federal income tax payments for that year were $108 million. That works out at a Federal income tax rate of 10.8 percent of their billion-dollar income.

Mobil tried to fool all of us by including in their tax "payments" $327 million in "deferred" taxes—the taxes they didn't pay because of special tax-shelter provisions such as intangible drilling cost deductions and accelerated depreciation. These "taxes," once deferred, are never paid because each succeeding years brings more of these tax breaks. So Mobil paid 41.9 percent in taxes only if you count $327 million they didn't really pay.

Mobil continues this flimflam with the claim that their overall tax rate, at home and abroad, was 77 percent in 1977. This figure includes not only deferred taxes but also large royalties paid to the OPEC nations and disguised as income taxes so they qualify for the foreign tax credit. These "taxes" are really the cost of buying the oil from the OPEC countries, but each dollar the companies pay to OPEC reduces their U.S. taxes by a dollar.

Mobil also claims that their profitability is below the average for U.S. manufacturing industries. Actually, Mobil's profit margin (return on equity) for 1979 is running at an annual rate of 20 percent—almost twice that of the company's 10.9 percent return before the oil embargo and above the manufacturing industry average of 18 percent. Its profits for the first quarter, in fact, were up by 82 percent over 1978's first quarter.

In another ad, Mobil argued they should receive the OPEC cartel price for newly discovered U.S. oil. They said they only want the same price that foreign producers get for their oil. They misspoke again. Foreign producers only keep about 5 percent of the price of OPEC oil. The OPEC nations demand royalties and excise taxes and a share of production profits equal to 95 percent of the world price. Mobil wants to receive the world price but keep most of it for itself.

Mobil may believe that they can say what they want when they spend millions from their record profits on these ad campaigns. They are not chilling the debate. They have frozen it with further distortions in an attempt to convince all of us that they need more of our hard-earned money.

The final salvo was fired at Mobil by Kenneth Flamm, a research associate, and Kenneth Oye, a guest scholar, both at the Brookings Institution. They wrote on September 18:

In his letter of July 31, James Q. Riordan of the Mobil Corporation argued that Mobil's effective 1977 U.S. income tax burden was 41.9 percent of pretax earnings, well above the Eagleton-Vanik estimate of 11.1 percent. Mr. Riordan's figure leaves the reader with the impression that Mobil is staggering under a massive tax burden. Quite the opposite is true.

As Mr. Riordan observed, the difference between these two figures is attributable to the definition of "taxes." The Eagleton-Vanik figure was based on current tax provisions alone. The Riordan estimate was based on both current and deferred tax provisions, in accordance with Mobil's accounting conventions. Roughly three-quarters of Mobil's 1977 tax provisions were in fact deferred taxes—tax liabilities accrued in 1977 that will not fall due until future years.

Just how much of an economic burden are such deferred taxes?

First, inflation erodes the real cost of deferred taxes. At 10 percent inflation, a dollar of deferred tax liability payable in 10 years is equivalent to a 39-cent tax liability payable today.

Second, the deferred taxes amount to an interest-free loan from the taxpayer to Mobil. If Mobil can invest in an asset yielding a 5 percent after-tax real rate of return over a 10-year period of deferral, that dollar of liability would be further reduced to the equivalent of 25 cents payable today. As a rule of thumb, the longer the deferral, the lighter the real tax burden.

In practice, it is difficult to calculate the real present value to Mobil of its deferred tax liabilities. The amounts and years in which deferred liabilities will fall due cannot be predicted. Furthermore, it is difficult to forecast the inflation rate and the corporation's probable real rate of return.

It is clear, however, that Mobil's deferred taxes are paid only after a substantial delay.

In 1977, Mobil accrued an additional $420.7 million in deferred tax liabilities, bringing its stock of deferred tax liabilities (U.S. and foreign) up to $1.3 billion. Information on Mobil's S.E.C. Form 10 indicates that approximately $6.3 million of deferred tax liabilities became payable in 1977. On a "flow-through" basis—comparing Mr. Riordan's figure on *actual pretax U.S. earnings* ($1.038 billion) with *actual tax payable* toward both current ($107.5 million) and deferred

($6.3 million) liabilities in 1977—Mobil's 1977 effective U.S. income tax burden was no more than 11 percent.

Mobil is using tax breaks approved by the Congress in a perfectly legal manner. It is patently misleading, however, for Mobil's officers to attempt to obscure the magnitude of those benefits, by trying to slip them under the rug of accounting conventions.

This debate demonstrates the problem confronting the big oil companies. Despite the millions spent on public relations, many people, including some very influential ones, just don't believe them. Nevertheless, the industry valiantly continued its efforts to become better understood even as the career of its leading public relations man took two surprising detours.

In November 1979, Herb Schmertz announced he was taking a six-week leave of absence without pay to organize advertising, polling, and media relations for Senator Edward Kennedy's campaign for the Democratic presidential nomination. There aren't many liberal Democrats in the upper ranks of Big Oil, although Schmertz was always a believer in the free enterprise system he was paid to defend.

Connoisseurs of black humor relished the fact that the architect of Mobil's public relations campaign had gone to work for one of the oil game's harshest critics. For Senator Kennedy is against just about everything that Mobil advocates. He has urged a stiff windfall profits tax, fought to retain controls on domestic oil prices, and backed legislation to curb acquisitions outside the oil industry by oil companies, like Mobil's takeover of Montgomery Ward. Schmertz, however, had a long-time relationship with the Kennedys and had worked in the campaigns of Senator Kennedy's brothers. Since he would not be involved in the formation of energy policy in the Kennedy campaign, he said that he saw no conflict of interest in his temporary defection. Others were not so sanguine. A spokesman for another big oil company told the Associated Press: "We are stunned. Whenever Mobil was attacked, Schmertz wrote knee-jerk ads that sounded a little to the right of Attila. Now he is joining a candidate who is a little to the left of center, to say the least. It seems to me that you either believe in what you're doing, or you don't." Jody Powell, press secretary to President Carter, commented: "I hope that Mr. Schmertz is as successful with Senator Kennedy's image and credibility as he has been with the major oil companies."

As it turned out, Schmertz's expertise did not help Senator Kennedy win the nomination. He returned to mastermind Mobil's continuing public relations effort, but did have another surprise to offer. Somehow he had found the time to write a novel with Larry Woods, another Mobil vice president. Titled *Takeover*, the book was published by Simon and Schuster in January 1980. Its plot involves a business acquisition and features much sex, bribery, corruption, blackmail, corporate intrigue, and, the jacket blurb promises, "a lot of other things people don't learn in business school." The blurb does not identify the authors as vice presidents of Mobil and Schmertz has said that the book has no relationship to Mobil or the oil game and is only an "interesting story about some people who engage in a caper."

Sales of the novel were reportedly light and most reviews ranged from snide to scathing. The *New York Times* observed:

> Wait a minute! Do we have Herb Schmertz, the omnipresent spokesman for Big Oil, saying—at least implying—that businessmen dump their morals in the garbage to get another rung up the ladder? . . . Or, then again, is he just following the tried and tested path of numberless commercial novelists, that is, dump anything into the book. It doesn't have to be real; it just has to titillate.

The review in the *Wall Street Journal* by David Sanford, managing editor of *Harper's*, was tougher on the novel than Herb Schmertz had ever been on his industry's critics:

> Mr. Schmertz, if not Mr. Woods, is much too famous a man to have any right to expect that book-jacket obscurity would keep him and Mobil Oil from being linked. The name "Herb Schmertz" is to the flack's calling what "Uriah Heep" is to false humility and "Benedict Arnold" is to traitors. Herb Schmertz is

> a major reason PBS is known as the Petroleum Broadcasting System. He brought "Upstairs, Downstairs" and other "quality" British imports to American television. He thus has been called (by *Esquire*) "a major force in our cultural life" and (by Sally Quinn) "the most powerful, most successful public relations man in America."
>
> It is Herb Schmertz who buys space in newspapers to propagate Mobil's social philosophy. It is he who cultivates Congress. For his efforts he is paid a nice salary, something more than $200,000 a year. So why is this man writing bad dirty novels that could only embarrass the boss? . . .
>
> "Takeover" is dreadful on so many levels that it is almost possible to enjoy, but it is impossible to admire. Businessmen as depicted in the book are a thousand times more mercenary and avaricious than even Ralph Nader would ever dream to claim. The undertaking thus does not serve business or the truth about how business is conducted in America.

Mr. Sanford concluded his review by offering the opinion that the oil men novelists "give new meaning to the word 'crude.'" This may be a little harsh on the man who brought you *Masterpiece Theater*. Perhaps one day Herb Schmertz will write a factual book about the basic reason for America's energy problem. For such a reason exists.

Chapter Seven

Getting to the Bottom of the Gasoline Barrel

Gasoline is the root cause of America's petroleum woes. It accounts for the largest share of the total petroleum America burns away at 43.1 percent (the next biggest share goes to heating oil and other distillates at 17 percent). Americans, with less than 5 percent of the world's remaining proven reserves of the crude oil from which gasoline is refined, own almost 40 percent of the world's passenger cars and consume almost 60 percent of the world's gasoline at a rate of more than 100 billion gallons a year.

For all their addiction to demon gasoline and the internal-combustion-powered automobile, Americans know as little about how the supply and price of gasoline are arrived at as they do about offshore drilling in the Beaufort Sea. Yet you don't have to be a Nobel laureate in economics to understand what's going on. It is only necessary to accept two basic premises: One, Americans use far too much gasoline. Two, the oil business is in business to make money and oil men are just as human as the rest of us in their desire to make a lot of it.

The logical place to begin to understand what's going on is at the local gasoline station—about the only place where the petroleum industry and the public it serves make direct contact. How do these businesses at the lowest but most visible end of the exploration, production, transportation, refining, and marketing funnel function? Why did they have to close down for lack of supplies in the spring and summer of 1979, and why were they then overflowing with gasoline—at higher prices, of course—just a year later?

The first step toward understanding the role of the gasoline service station is to realize that the oil companies do not actually own and operate most of the stations that sell their branded products. Fewer than 10 percent of the stations are owned by oil companies and staffed by their salaried employees. The rest are operated by that American folk hero, the independent businessman. He either owns the station himself or leases it from a company.

Under license, the oil company then acts as the sole supplier to the independent businessman, who tacks on his own profit after paying for the supplies. So to get mad at Joe of Joe's Texaco Station when you run into shortages or high prices is to get mad at the wrong target. Joe takes what he's allocated by Texaco under government regulation and is told what it will cost him. His own profit margin is controlled—or is supposed to be—by the government.

Once that basic arrangement is understood, the world of the retailer becomes more Byzantine. There were 226,000 retail gasoline outlets in the United States in 1972. Seven years later, their number had shrunk to less that 167,000. The reasons for this decline offer an interesting, if rather cold-blooded, example of the free market system at work.

Traditionally, the large full-service stations selling major brands charged more than their smaller nonbranded competitors. They had more overhead (electricity, heat, insurance, salaries for pump attendants, etc.) and they offered a wide range of tempting services (wipe your windshield? engine tune-up? change your oil or snow tires?). They also had one or more salaried mechanics in their service bays for major repairs.

The operators of such full-service stations—which still make up 70 percent of the retail market—earned two-thirds of their profits from service and repairs and only one-third from selling gasoline. The latter was a lure to draw in customers for the more profitable part of the business. The oil companies, too, did not make their biggest profits from marketing—or downstream—operations but from their production—or upstream—operations.

This comfortably stable relationship was given a severe wrench by the shortages and the entry into the market in force of independents selling nothing but nonbranded gasoline, often on a self-service basis. These independents, selling gasoline with obscure and catchy names (Merit, Fill Em Fast, Old Colony, etc.), got their supplies from the oil jobbers, independent brokers, and marketers. The latter bought them from small refineries and on the spot market where, when gasoline is plentiful, it is cheaper than the product sold under long-term contract by the majors. They also bought the leftovers from the major refineries, which are cheap when supplies are ample. Motorists who worry about the independents selling bottom-of-the-tank or watered-down gasoline have no cause for concern. Branded names like Exxon or Mobil or Texaco set a standard of quality, but it's common practice for independents to buy from the majors and even for the majors to make exchanges with one another. But because of their cheaper supplies and lower overhead, the independents were able to undersell the major brands.

The majors didn't stand still for that for very long. Their executives reasoned that with supplies tight and prices high, fewer stations selling more gas per pump spelled higher profits. High volume became the goal as part of a fundamental change that has come over the oil business in recent years.

When the multinationals began to lose control of cheap foreign oil to the producer countries and Washington moved to reduce or eliminate hefty tax writeoffs like foreign tax credits and the percentage depletion allowance, the traditional profitability of the production end of the business became less pronounced. That forced the majors to compete with their smaller rivals in marketing on the latter's terms.

The first step was to get rid of as many low-volume stations as possible, especially in rural areas. This meant the beginning of the end for another American folk symbol—the friendly neighborhood station where you could stop to chat with the owner about the Red Sox or deer hunting, cash a check or buy a bag of fertilizer, and pick up a container of coffee at the diner next door across from the deserted train depot. If it also meant sacrificing the smaller dealers, many of whom had been in business for twenty years or more, and their faithful rural customers to cost efficiency and mass merchandising, that is the way of the world in the profit-oriented oil business.

Texaco, for example, in its 1978 annual report said that it was acting "to improve the economies of its downstream operations." In plain English, that meant it was closing marginally profitable retail outlets. Since 1973, Texaco reported, it had eliminated some four thousand of what it called "high-cost investment retail outlets." In March 1979, the company announced that it was closing another two thousand stations in the Middle West and upper New York State. The rule of thumb for all the big companies is that if a small station isn't making a good profit, the company isn't either.

The majors' next step was to undertake marketing innovations at their remaining high-volume stations, such as the hybrid station equipped with mechanics and service bays but offering only self-service gas, and the "split-island" station with both full-service and self-service lanes. Finally, the majors were not above marketing their gasoline under the obscure or catchy brand names associated with the smaller independents. When motorists buy Hi-Val, Sello, Reelo, and Big-Bi, they are really buying Mobil. Alert is actually Exxon, while Go-Lo, Economy, and EZ-Go are Gulf.

With these moves, the majors were able to retain their dominant role in the domestic gasoline market, where they were in the enviable position of being able to sell everything they produced. In the typical year of 1978, the ten largest marketers held a 55.72 percent share of all gasoline sales, as follows:

Shell	7.71%
Amoco	7.60
Exxon	7.30
Texaco	6.82

Gulf	5.86
Mobil	5.50
Standard Oil of California (Chevron)	4.72
Atlantic Richfield (Arco)	3.76
Union Oil	3.27
Sun Oil	3.18

How are prices at the pump arrived at, and why do they keep rising? It all begins at one of the some three thousand refineries in the United States. A refinery can be a "teakettle" plant belonging to an independent capable of processing 150 barrels of crude oil daily, or a complex giant like Shell's Norco Refinery in the swamp-and-bayou country 25 miles west of New Orleans with a daily capacity of more than 450,000 barrels of crude. The crude is pumped into the refinery through pipelines in a silent, continuous stream, creeping like molasses or flowing like water, depending on its type. The one preferred for refining into gasoline is sweet crude, which is light and brownish and has a sulfur content below 1 percent. Sour crude contains over 1 percent of sulfur and other mineral impurities.

At the refinery, this mixture of thousands of different hydrocarbons is converted to a fine spray. It boils, bubbles, and separates in a maze of chimneylike columns and pipes and its molecules are cracked apart and recombined in towering cylindrical vessels. The end result can be gasoline, home heating oil, diesel fuel, jet fuel, petrochemicals, or any one of hundreds of other petroleum products.

Therein lies a dilemma that does not exist when the supply of crude is plentiful. When it is tight, a worrisome decision on the refinery's mixture of products must be made. Should more of the available crude be processed into gasoline, risking a shortage of heating oil in the winter and the wrath of homeowners, or should the concentration be on heating oil, risking gasoline shortages and the outrage of motorists in spring and summer?

If the decision is to tilt in favor of gasoline, there is still another, longer-range problem. Unless millions of dollars are spent to expand the sections of existing refineries that process unleaded gasoline to boost its octane rating, there may not be enough to go around in two or three years, regardless of the crude supply. Spending these immense sums will be justified if the demand for unleaded, high-octane fuels keeps rising. If it tails off as light, more fuel-efficient cars come on the road, the oil companies will be left holding the bag of serious damage to their bottom lines.

This quandary is the fruit of the federal government's decision to achieve both gasoline economy and environmental protection. In 1975, responding to environmentalists' insistence that the lead in automobile exhausts is one of the prime poisons in air pollution (still a matter of dispute), the government decreed that all new cars must be manufactured to run on unleaded fuel. This legislation certainly had good intentions, but like many another program undertaken with good intentions, it resulted in unforeseen problems.

First, all gasolines are not alike, and the most crucial difference among them is their octane rating. You don't have to be a chemical engineer to understand that octane is not a dynamic factor that adds power. It does nothing but depress the tendency of an engine to knock. Different engines require octane with a different rating. If it's too low for a particular engine, instead of burning smoothly, the fuel-air mix will explode and the engine will knock or "ping." This can cause as much internal damage to an engine as hitting it with a hammer, as well as poor performance. It also wastes gasoline. General Motors solved this problem back in 1923 by developing the antiknock additive tetraethyl lead (TEL). It was blended with the gasoline that was cracked from crude oil at an octane level of about 60 percent. Tetraethyl lead raised the octane level to over 95 percent. Engines performed smoothly until 1975, when lead-free fuel was mandated for all new cars.

Meanwhile, the auto industry came up with the catalytic converter that successfully eliminated pollutants. But this ingenious solution created yet another problem. If such emission-control systems are to work properly, there can be no lead in the gasoline fed into them. Lead destroys catalytic converters. But when lead is removed, the octane rating drops. This, in turn, subjects new automobile engines to knocking and to serious damage. Frantically, the oil companies searched for new additives to replace TEL. They have found none acceptable to the government.

At present, there is one safe method to boost the octane level of unleaded gasoline. That is to step up the refining of crude oil by two processes called alkylation and catalytic reforming. Well, good, you might think with relief, there's a technological solution. Unfortunately, there's a catch (as always). The two processes are not only costly but also devour more crude. For every 10 barrels of leaded gasoline, refineries get 9 barrels of unleaded from the same amount of crude. The obvious result is a further drain on America's crude supplies, as well as on the motorist's finances since premium unleaded and unleaded cost more to process and the oil companies pass on those costs to the consumer.

For motorists, the confusing octane rating question can be answered simply. Older cars without catalytic converters can run smoothly on any kind of leaded gasoline. Of course, they will get poor mileage, might cause pollution, and will not last forever. But brand names and higher prices don't mean better leaded gasoline and better performance.

The same holds true for the premium unleaded and unleaded gasoline required by post-1975 cars. Motorists should shop around for the cheapest gasoline they can find that their car will burn without knocking. Some have chosen a different option in a form of civil disobedience by using leaded gasoline in new cars, which destroys their catalytic converters and causes even more pollution than older cars. Others have had body shops remove the converters entirely or cut larger openings in the filler neck. In 1979, the Environmental Protection Agency estimated that as many as 15 percent of motorists engaged in this fuel switching. They can't be prosecuted, although service station operators who do the illegal work can be fined up to $10,000. The catch is that the EPA doesn't have enough inspectors to enforce the law.

The question of octane ratings is more complicated for the oil companies, whose basic concern is profit. They don't object to the phasing out of leaded gasoline, for their profits on the higher-priced unleaded product are quite satisfactory. But they don't want to expand the facilities that process unleaded, only to have demand for it fall off, putting them over a financial barrel, so to speak.

Leaving the resolution of this problem to be wrestled with in the boardrooms of Big Oil and the chambers of the government agencies that are supposed to regulate it, let us return to the service station and the question of pricing. It still begins with the crude oil that goes into the refinery, whether it be domestic or imported. When that crude oil costs more, the gasoline processed from it has got to cost more. Let us follow the hegira of crude oil imported from Saudi Arabia in 1979.

The cost of drilling, pumping, and transporting crude from the Saudi Arabian deserts to Persian Gulf ports was 30 cents per barrel, or less than 1 cent per gallon. However, by the time that 1-cent-a-gallon crude was shipped, refined, and sold as gasoline to American motorists, the price had soared more than 7,000 percent.

What accounts for such a dizzying rise? First, the royalties, fees, and taxes of the Saudi government boosted the price per barrel to $14.55, or 35 cents per gallon. Tanker costs and profits for their operators added about $1.25 per barrel, or 3 cents per gallon. U.S. refineries, after getting a refund from the federal government's complex entitlements program, wound up paying about $14.30, or 34 cents per gallon, for the Saudi crude.

The costs of refining, wholesaling, marketing (including advertising), and transporting the gasoline right to the pump resulted in the oil companies selling it for 54.4 cents per gallon, a markup of 20.4 cents. Federal, state, and local taxes added 13.2 cents per gallon. All these steps brought the cost of the originally 1-cent-per-gallon Saudi crude up to a national average of 77.3 cents per gallon in the spring of 1979.

That average cost had soared to well over $1 a gallon by the fall of 1980, chiefly because of the inexorable climb of OPEC crude oil prices. Dollar-a-gallon gasoline breaks down like this: 43

cents for the oil company's costs for crude oil; 20.5 cents for the oil company's refining costs and profits; 6 cents for middlemen's costs and profits; 17 cents for federal, state, and local taxes; 13.5 cents for retailers' costs and profits.

The big oil companies protest that they are not profiteering. Their published claims for profits per gallon range from Texaco's 2.5 cents to Shell's 4.1 cents to Chevron's 5.1 cents. Unmentioned in their apologia is the fact that those seemingly modest profit margins produce awesome overall profits because of the volume of gasoline sold. Between January and July 1979, U.S. gasoline profits doubled from $20 million to $40 million a day.

Have service station operators profited from this bonanza? They claim, like the oil companies that supply them, that they really aren't making enough money. Then why have price increases at the pump outstripped OPEC price hikes, and why do pump prices vary so widely? The answers can be found in the laws of supply and demand and the vagaries of government regulation, and to a lesser degree in the human penchant for greed. Let us examine how all these factors interacted during 1979, a year of gasoline shortages and price rises that is destined to be repeated.

To prevent price gouging, the Department of Energy adopted a formula in 1974 in response to the shortfall of petroleum supplies created by the Arab oil embargo. Under it, a station operator who, for example, made 9 cents a gallon during the base period of May 15, 1973, could still make only 9 cents a gallon in 1979, with allowances for increases in wholesale price, rent, utilities, and environmental costs.

Meanwhile, the major companies benefited from a regulation known as "tilt." Since gasoline costs more to produce than most other oil products, refiners were allowed to "tilt" more of their profits into gasoline. This increased the price to the dealer and ultimately to the consumer by about 4 cents a gallon.

Also built into this well-intentioned price control system was a mechanism called "banking." Market pressures kept prices below the federal ceilings when gas remained plentiful. The refineries of the major companies and their dealers could then "bank" or save price increases they were entitled to but could not apply when demand was weak and they were vying for customers. They could then legally reapply that "bank" to the price they were allowed to charge when supplies and competition diminished, as happened after the Iranian cutoff of early 1979.

By some incredible oversight, no one in the federal government, or specifically the Department of Energy, had foreseen the end result of the "bank" mechanism. During the shortages that took place in the spring and summer of 1979, dealers could charge the legal maximum limit for gasoline. They could then legally add to that price the amount they had "banked"—or not been able to charge—when gasoline was plentiful and they were competing with other dealers for business. Thus motorists found themselves paying the controlled ceiling price, and over and above that paying back every penny they had saved earlier—a total increase that outstripped the OPEC price rise. Incongruously, this price-control system had precisely the same effect as no price controls.

The question of why prices varied so widely from region to region and even block to block can be answered simply. There was no national regulation that made profits a fixed percentage of the wholesale price. Individual dealers were left to figure out their own prices and profits based on their mid–May 1973 operations, of which many had not kept accurate records. These figures, of course, varied widely, as did transportation costs, local gasoline supplies relative to demand, and differences in costs for rent, labor, and utilities.

Finally, many gas stations simply ignored the legal prices. The Department of Energy estimated that half the gas was sold above the official ceiling, or at what the market would bear. At a few stations in Boston and New York City, the price hit $1.70 a gallon. The Assembly Oversight and Investigation Committee of the state of New York reported in September that 64 percent of the stations it surveyed in New York City were charging motorists more than the allowable 15.4 cents a gallon, plus taxes, above the oil company's price to the dealer. The committee also accused the Department of Energy of "a shameful cover-up of its own ineptitude in obtaining compliance."

Price gouging was punishable by a fine of $10,000. The catch was that black marketers vastly outnumbered Department of Energy inspectors. There were some 167,000 gas stations and fewer than 50 DOE auditors assigned to them. The DOE's Office of Special Counsel for Compliance, set up in 1977 to uncover profiteering, had 400 auditors and 200 lawyers, but their attentions were largely focused on the major oil companies.

Despite this surveillance, it is impossible to pinpoint how much the major companies profited from the price rises and shortages. More than five years after the 1973–1974 crisis, the DOE had failed to complete audits of the companies' pricing policies beyond the year 1976. However, in early December 1979, the DOE did charge Mobil with pricing violations of $500 million from 1973 to 1976. That brought the total of alleged overcharges charged in complaints against Mobil to more than $1 billion. Similar charges were filed against Gulf, Amerada Hess, and Chevron, a subsidiary of Standard Oil of California.

Shortly thereafter, notices of probable violation, which carried no charge of willful wrongdoing, were given to Shell, Atlantic Richfield (Arco), Texaco, Conoco, Marathon, Standard Oil of California, and Standard Oil of Ohio (Sohio). The allegations concerned not just the pricing of gasoline, but also of home heating oil, diesel fuel, propane, heavy fuel oil, jet fuel, and kerosene. They involved more than $1 billion worth of probable violations of federal price rules and regulations. DOE officials estimated that the full amount of alleged overcharges would come to $25 billion when the department's investigation, one of the most detailed in federal history, was finally completed. That figure makes the few cents a gallon the service station operator made if he engaged in a little price gouging seem like very small change indeed.

The accused companies were quick to respond. William Magee, a vice president of Arco, commented that the allegations "represent a continuation of the special counsel's politically motivated program of claiming headlines with big numbers, but little substance." Bruce McFall, a vice president of Conoco, protested that the DOE's accusation against his company was a "patchwork of assumptions, guesses and projections."

The companies denied any wrongdoing. A Mobil spokesman in New York City commented: "If there have been errors, they were predominantly those of D.O.E. and its predecessors in drafting the regulations poorly, in foot-dragging when we have requested explanations, in issuing rulings and interpretations that differed from the initial regulations, and then in attempting to enforce those rulings retroactively."

Meanwhile, the DOE's legal pipeline became clogged with a file of about 160 allegations of violations of federal regulations totaling $9 billion against the thirty-five largest American oil refiners. The industry contended that the DOE regulations were contradictory and confusing and that the agency engaged in retroactive rule making. It was estimated that it would take years of legal wrangling to determine how much fire there was to all this smoke. However, by December, the DOE had won consent decrees amounting to $660 million from Mobil, Phillips, Gulf, Kerr-McGee, Cities Service, and other companies. Consent decrees are not convictions but "settlements" in which the companies that sign them neither admit nor deny wrongdoing, but agree to stop what they have been doing and make a financial settlement.

Getty Oil, which had been accused by the DOE of violating federal price regulations on crude oil, natural gas liquids, and petroleum products, chose to sign a consent decree and pay $75 million rather than fight the charges in court. The settlement included a novel provision in which Getty agreed to pay $25 million into an escrow account to be administered by the DOE to "provide relief to economically disadvantaged people in meeting their energy expenses for this winter." Thus the oil companies, which saw themselves as corporate Medicis promoting culture and the virtues of the free market, became sponsors of a social welfare program.

In February 1980, the DOE hit its biggest single target up to that time. Standard Oil Company of Indiana, one of the largest gasoline retailers (under the Amoco trade name), agreed to a $690 million consent decree settling all alleged pricing violations since 1973. As in earlier settlements, the DOE did not spell out its allegations against Indiana Standard. Nor did the consent order constitute an admission or finding of violation of government regulations by the company. And the oil companies went right on complaining that they hadn't done anything wrong and still weren't making enough money.

If Indiana Standard (and the others) felt that it was innocent of any overcharging, why did it agree to a $690 million settlement? From its corporate headquarters in Chicago came the explanation that the company "was willing to enter the agreement with the Department of Energy to avoid further disruption of its business activities and the continuing expense of protracted, complex litigation."

To avoid this bothersome disruption and expense (its earnings had increased 70 percent in the last quarter of 1979 and 40 percent over the whole year), Indiana Standard specifically agreed to do the following, according to a report in the *New York Times:*

- Pay $29 million directly to reimburse past buyers of home heating oil and diesel fuel, mainly public utilities and bus lines.
- Pay $71 million into a Federal escrow account—one entrusted to a third party until certain conditions are met—for disbursement to persons suffering the most from rising energy bills, mainly low-income families.
- Forgo $180 million in price increases for producing gasoline and propane that it could have legally imposed to meet rising costs.
- Reduce product prices 2 cents a gallon to some customers over the next 18 months at a cost of about $25 million.
- Spend $105 million to modernize a Texas City refinery so that it can process heavy oils and crudes high in sulfur content, thus accelerating gasoline production.
- Spend $178 million to accelerate exploration for domestic oil supplies in an effort to reduce imports.
- Spend $128 million to accelerate the production of oil from old fields that are only marginally profitable.

It will be noted that in the largest of these reimbursements Amoco is really paying itself by investing in projects that will later return a profit to the company. None of the money will reach the pocketbooks of the individuals who may have been overcharged for Amoco gasoline or heating oil. As DOE Special Counsel Paul Bloom explained, it would be impossible to locate these millions of customers.

Consumer groups were unhappy with this agreement. They considered a settlement of $690 million a slap on the wrist in view of oil company profits. "As usual," said Ellen Berman, executive director of the Washington-based Consumer Energy Council of America, "the oil industry has come out on top, showing it pays to violate the law."

The man responsible for enforcing the law the industry stoutly maintained was contradictory, confusing, and capricious was Paul Bloom. In 1977, Bloom was appointed special counsel to the Energy Department to oversee enforcement actions aimed at ensuring that the petroleum industry obeyed federal rules and regulations covering pricing of petroleum products. In short, he was the federal government's instrument for protecting the public from the alleged excesses of the oil companies. It was an enormous job, considering that Bloom and his staff of six hundred lawyers and auditors had to investigate not just the major oil companies, but some two hundred refiners, nineteen thousand crude oil producers and twenty-five thousand wholesalers.

In 1980, Bloom was forty years old. He was a broad-shouldered man with a cowpoke mustache who came to his high-ceilinged, wood-paneled office in the old Post Office Department from the adobe haciendas of Sante Fe, New Mexico, his home, where he had spent twelve years as a natural resources lawyer for the state. It was, of course, quite a jump from that modest job to jousting with some of the captains of American industry and what Bloom has called the "legal gun slingers they send against me" in multibillion-dollar litigation. To the oil business, he became quite possibly the most unpopular man in America. The industry contends that DOE regulations were unclear, that interpretations were either contradictory or unavailable, that controls were needless. As Bloom has said, they "bitterly fought attempts to enforce them."

On the other hand, some consumer groups have attacked

Bloom for not extracting sufficiently tough settlements from the oil companies and for failing to refer criminal charges against them to the Justice Department. It is true that the special counsel's allegations were not criminal in nature, but civil charges involving technical violations. No conspiracy was alleged, no willfulness asserted.

Paul Bloom, viewed as a fanatical enforcer by corporate oil and a pushover by the consumer groups, was probably the best qualified man in America to tell angry motorists whether or not they were being ripped off by the oil companies. Although Bloom did provide some insights in an interview with the *Los Angeles Times*'s Martin Baron, which was printed in its issue of February 10, 1980, he had no easy answer that would stand up in a court of law. Some of the questions and answers from that interview follow:

Q: Oil companies say many energy regulations were vague and that their requests for interpretations often went unanswered. How can you now say they violated the rules and penalize them for it?

A: Am I supposed to sit back and say, just because you didn't get answered, you have a license to overcharge your customers by $80 million? I mean, it doesn't follow. What follows is a criticism of an inept system in terms of not providing a quick response. But it is not a defense against an unlawful interpretation of the price control regulations.

It's not a soccer game we're playing between the company and the government. The government is supposed to be the referee in the price control system between the company and their customers.

Q: Did the oil companies purposely violate the regulations or did they simply interpret the rules differently?

A: Some took the position: let's figure out what's the greediest call and what's the most arch-conservative, talk to our lawyers about our relative liability and make sure neither would be a criminal possibility. And then we take a middle course or take something less than the greediest course and, at the same time, try to get an interpretation from the agency. . . .

That might expose them to some non-zero liability, but it is by no means the most egregious corporate behavior. Some, on the other hand, could not have been better corporate citizens. . . .

Q: Yet many newspaper stories and columns describe you as an energy cop going after oil company price gouging and windfall profits.

A: I can show you newspapers that used ripoff in headlines when I stand up at briefings and say: civil not criminal; no willfulness asserted; only charges, no adjudication; total dollars not overcharges; cost challenges are not overcharge dollars. And I get up and say the total we announce today is $600 million, and the next morning I will read in some papers, like clockwork, "Government charges $600 million ripoff."

Mobil bitches about this. The problem is they bitch at the wrong people. And they use it for their own tendentious purposes. But [Mobil Chairman] Rawleigh Warner is not wrong to feel hot under the collar about being accused of a ripoff for civil violations. But I never accused them of that. . . .

Go and ask any cab driver in Washington who's behind the so-called fake energy crisis, and he'll say, "Those sons-of-bitches running oil companies, they're stealing us blind." The country is full of people who don't trust or like oil companies. I didn't invent that. I don't inflame it. I live with it and so do they. They occasionally act like I invented it. That ain't hardly fair. . . .

Q: Oil company lawyers say they aren't even close to settling the charges. They say you're proposing such unreasonable terms that there will be few more settlements.

A: You're being hustled. It's a little exaggeration to say they're lined up at the door. Nobody's lined up holding checkbooks. Half a dozen more companies are talking to me. . . . They have to look aloof. They have to look relatively unconcerned. They have to look macho. They have to look like they're hiding behind the biggest, toughest, meanest energy lawyers in the west. . . . On the other hand, I'm in a position to know, as you are not, how many companies are talking to us, how much larger the numbers in those settlements will be than the ones I've previously announced. Therefore, it is easy for me to be somewhat relaxed about the comments that you recount. In fact, slightly amused.

So we have a somewhat relaxed and slightly amused DOE special counsel holding that the oil companies were not engaged in a criminal conspiracy to rip off the public, but nonetheless could be coerced by some sort of evidence into agreeing to handsome settlements out of court. The episode left a confused public still wondering if the oil companies had done anything *wrong*.

One government official who was convinced that the oil companies were doing something wrong was the president of the United States. On March 28, 1980, in a speech to the National Conference of State Legislatures, Jimmy Carter charged that Mobil had violated the administration's voluntary price guidelines by $45 million and castigated the company for being unwilling to refund that amount to motorists through a temporary reduction of 3 cents a gallon on its gasoline.

Mobil, which during an earlier disagreement with Mr. Carter's views had heard itself termed the "most irresponsible company in America" by the president, denounced the president's charge as "politically motivated" (1980 was a presidential election year). The corporation rushed into print to inform Mr. Carter that he had been misled in one of its detailed editorial-type advertisements, which ended on an injured note:

> In summary, Mr. President, the rules of the game were changed retroactively. We were not out of compliance under the rules in effect when we were selling the products in question. Anybody can be thrown into violation when rules are made retroactively. We oppose retroactive rule changes which put companies into violation, particularly when they are applied selectively against companies that speak out.

A compromise was reached in April 1980 when Mobil, still maintaining that it had acted in good faith, agreed to forgo $30 million in permissible price increases to offset the $45 million in alleged overcharges. But anyone who thought that the government was at last making substantial progress against oil company overcharges could derive slight comfort from a dispute that became public during a Senate Judiciary Committee hearing in early April 1980.

At this hearing, officials of the Justice Department charged that the enforcement of oil pricing regulations had been severely weakened by arguments over the division of litigating authority between the Justice and Energy departments, and that the latter's lawyers had often refused to cooperate with Justice Department lawyers preparing civil cases. Paul Bloom did not testify at the hearing, but said in an interview that he was "saddened and regretful that the Justice Department was not as flexible and responsive" as he had hoped it would be when he helped work out an agreement dividing legal responsibilities between the two agencies. Obviously, it would be difficult, if not impossible, to determine whether the oil companies were profiteering until Justice and Energy resolved their interagency conflict. Even then, no judgment would come until government lawyers completed a lengthy period of jousting with the imposing phalanx of legal talent employed by the oil companies.

As for the low man on the oil business totem pole, it did not appear that the service station operator was getting rich at the public's expense. A handful of high-volume gas-and-go stations selling nonbranded gasoline showed handsome profits. So did some branded dealers, like Mike Greenberg's Triple Gee Amoco station in northeast Washington, D.C. In July 1979, Triple Gee pumped 90,000 gallons and earned a profit of $11,429, an increase of $4,241 a month, or 59 percent, over January. But even with the "bank" provision and a little gouging, the average neighborhood station operator pumping 30,000 gallons a month did not get rich. With his supplies reduced, his worries increased by the allocation system worked out between the government and the oil companies, his profit margins held to the mid–May 1973 figure, and nearly half of that wiped out by inflation, he was—just barely—an economic survivor. Matthew Troy, executive director of the Long Island Gasoline Retailers Association, estimated that the average operator made about $20,000 a year, which sounds like a plausible ball park figure.

So, we are back to the exasperating question we began with: Who was responsible for the stiff price rises and would they at last taper off?

As 1979 slouched toward its end, four key oil producers —Saudi Arabia, Venezuela, the United Arab Emirates, and Qatar—as a sort of surprise Christmas present for American consumers, raised their prices by up to a third, the largest single increase in five years. They made this startling move four days before all the OPEC members were to sit down in Caracas on December 17 for their fourth price-raising session in a year.

The sparsely populated desert kingdom of the House of Saud, OPEC's biggest exporter and long regarded as its quintessential moderate, increased its crude oil price by $6 a barrel to $24, retroactive to November 1. The impact of the Saudi action was felt almost immediately by American consumers. Exxon, Mobil, Texaco, and Standard Oil Company of California, the four owners of Aramco, which pumps almost all of Saudi Arabia's oil, all announced rises in their wholesale prices.

In the New York City area, the rise in wholesale prices translated into an average jump of 4 cents for all grades of gasoline at the pump, bringing the average price to $1.20 a gallon. Not everyone believed that the Saudi action justified an immediate rise in oil company wholesale prices. Jerry Ferrara, executive director of the New Jersey Gasoline Retailers Association, representing three thousand service stations, said: "We don't agree with the major oil companies. We shouldn't have seen these increases until January." Agree or disagree, the dealers Mr. Ferrara represented had no choice but to pay the higher prices and pass them on to their customers.

Why did "moderate" Saudi Arabia raise its crude oil price by $6 a barrel, when it was already unable to spend its oil income as quickly as it earned it, and could gain no apparent benefit from precipitating a further erosion in the value of the dollar since it had extensive U.S. investments?

To give an idea of the magnitude of these investments, the International Business Machines Corporation, the world's biggest computer manufacturer, disclosed on December 27, 1979, that it had borrowed $300 million from the Saudi Arabian Monetary Agency, the kingdom's central bank. Almost five years before this deal, the American Telephone and Telegraph Company had borrowed $100 million from Saudi Arabia. The Saudis have also invested heavily in U.S. government securities.

Although unhappy with American support of Israel, Saudi Arabia has remained a "friend" of the United States since a World War II meeting between King Ibn Saud and President Franklin D. Roosevelt. The United States has provided a wide range of support to the Saudis, from hundreds of American technicians to multibillion-dollar arms sales. Saudi Arabia buys more American goods than either France or Italy. In reciprocation, the Saudis in 1979 maintained their production at 9.5 million barrels a day to avert shortages in the West and to restrain other OPEC members from driving up prices.

American diplomats and businessmen have long maintained that the conservative House of Saud—keeper of the Moslem holy places—provides an island of stability in an unstable region wracked by renewed Islamic fervor and anti-Americanism. (This was once an article of faith about Iran under the Shah.) Why, then, did the Saudis elect to lead a price increase by the so-called moderate producers, a move certain to harm the economies of their American and other Western "friends"?

To answer this question, we must play detective and try to penetrate a multinational industrial labyrinth unexampled in the modern world.

That highly secretive oil Goliath, the Delaware-based Arabian American Oil Company, is by far the world's largest oil-producing corporation. It produced some 9.5 million barrels per day in 1979 from the Croesus-rich fields of Saudi Arabia. Aramco's stock is not publicly traded, so it is not required to publish financial records. But financial analysts have estimated that in 1979 Aramco paid between $800 and $900 million to its four shareholders—Exxon, Mobil, Texaco, and Socal—and provided them with sizable tax benefits as well.

For example, Aramco paid Saudi Arabia the fixed price for the oil it produced and then collected a production fee of 25 cents per barrel. However, 85 percent of its payments were considered Saudi income taxes, which Aramco's four shareholders ultimately could use to reduce their U.S. income taxes. Each time

Saudi Arabia increased its crude oil prices, Aramco's local tax payments rose. And so did the benefits it reaped under the U.S. foreign tax credit.

In 1979, Saudi Arabia sold most of its oil to Aramco for $18 a barrel. Most other OPEC members were charging at least $23.50, while the price on the spot market was $36 or more. But Saudi leaders began to complain that Exxon, Mobil, Texaco, and Socal were not passing the "Saudi benefit" on to American consumers. Oil Minister Ahmed Zaki Yamani strongly suspected that the four companies were profiteering from Saudi "generosity" by diverting low-priced Saudi oil, which was intended to be refined in the United States, to refineries in Europe where there were no price controls, and then charging what the market would bear.

The Saudi Arabian leaders complained to William Miller, then secretary of the treasury, that the Aramco partners were reaping unwarranted profits from their low-priced oil during Miller's visit to Riyadh in November. "They feel the price they are selling at has not gone to the benefit of consumers and that it has been raked off by the oil companies," Miller told American reporters traveling with him, "and they are very upset about it. The Saudis feel that they have been taken advantage of by the oil companies." Miller speculated that the Saudis might raise their prices as a result.

The Aramco partners heatedly denied the charges. Privately, some of their executives labeled them spurious on the grounds that the Saudis owned 60 percent of Aramco and presumably know everything about its operations.

Two things can be said with certainty about the controversy. First, it highlights the divided loyalties of the Aramco partners. The giant multinationals have to walk a finer and finer line between the steadily diverging interests of producing and consuming states. In order to ensure their megaprofits and continued access to crude oil, they must avoid antagonizing their Saudi hosts or jeopardizing Aramco's concession (hence the private nature of the accusation that the Saudis' outrage on behalf of American consumers was spurious). Their profits do not depend on putting the interests of American consumers first.

Second, Exxon, Mobil, Texaco, and Chevron did in fact sell their gasoline, heating oil, and other products in the United States for a few cents less than their competitors' prices. But then came the reports of their third-quarter profits, which jumped by 73 percent to 211 percent and were ascribed by the companies themselves largely to overseas earnings. The revenue surge reportedly enraged the Saudis. Howard Kaufmann, president of Exxon, said there was no basis for this reaction:

> The fact that Saudi Arabia has been selling its crude for two dollars to four dollars a barrel below the price of other OPEC crudes this year has contributed to our refining and marketing margins in foreign markets. But some of this so-called Aramco advantage has been passed on to our customers in less aggressive prices. . . . United States customers have gotten full pass-through of lower costs.

Where does the truth lie? The Carter administration was forced to investigate the matter because of the politically sensitive nature of the Saudi accusation. Terence O'Rourke, legal aide on energy issues to Alfred Kahn, who was Carter's special adviser on inflation, wrote a study that charged that the few major American oil companies with access to foreign petroleum supplies had indeed diverted crude oil bought under contract to the higher-priced spot market and had increased markups to other domestic refiners:

> The bulk of foreign oil traded in international markets and imported into the United States is controlled by a handful of major international companies. Other companies buy all or most of their foreign oil from them.
>
> In recent months, as the world oil supplies became tight, these major companies reduced their third-party sales to other companies in order to meet their own needs and/or divert supplies to take advantage of high markups on remaining third-party sales.
>
> Their customers who were cut back were driven into very

> thin spot markets for oil where they bid up prices to extraordinary levels. They imported this oil at vastly inflated prices into the United States where it has had the double impact of driving up prices for both domestic crude oil and refined products.
>
> In sum, market conditions over the past several months offered companies with the greatest access to foreign crudes and the least dependence on it numerous opportunities to make unprecedented profits. The market also provided unusual opportunities for such companies to use their special access as leverage to boost their profits even more. The outcome, of course, can be seen in the recent reports of third-quarter earnings. These reports show that, while almost all oil companies did well, those with the greatest access to foreign crudes did best.

The oil companies denied O'Rourke's charges and there was widespread disagreement within the Carter administration itself over the accuracy of the Saudi views. "One thing is clear to us," said Deputy Secretary of Energy John Sawhill. "The four Aramco partners generally underprice their competitors in the United States. What isn't clear is whether they have done so to the full extent of the differential in their costs. Until we get more detailed audits, we don't really know."

The incredible fact is that the United States government really didn't know. Allowed to operate as sovereign states as befitted charter members of the Seven Sisters, Exxon, Mobil, Texaco, and Socal—after meeting their domestic refinery needs—could sell as much of their Aramco crude as they wanted to through their subsidiaries in the higher-priced markets of Europe and Japan. And the government not only had no hard evidence as to how much of this was going on, but even if it had had evidence, it could do nothing about it, for it had no jurisdiction over the foreign subsidiaries of the multinationals. There was no mechanism to compel Aramco to sell or stockpile *all* of its relatively cheap Saudi crude within the United States, thus averting shortages there and holding prices in check.

The administration was in the ludicrous position of basing its monitoring of the international movements of crude entirely on data supplied by the companies themselves. This situation was an extension of the federal government's total reliance on the American Petroleum Institute for such vital statistics as imports, domestic crude production, refinery output, stocks of crude oil and refined fuels. This dependence, which cost the government much of its public credibility as the regulator of the oil industry, was not abandoned until the end of 1979 when the DOE began to assemble its own data.

Meanwhile, the Saudi Arabian officials, many of whom had been educated in America (Sheik Yamani graduated from the Harvard Business School), could understand figures and dollar signs as well as anyone at the DOE, the American Petroleum Institute, or Exxon. Given the lower price they were charging their Aramco guests, which amounted to a $95-million-a-day subsidy to the West, and the subsequent exploding profits of these four Sisters, it is difficult to reproach the Saudis for jacking up their prices 33.3 percent before the OPEC meeting in Caracas even began.

In tracking this international detective story, it will be impossible to say "who done it" until the unlikely happens and the files of the Saudi Arabian government and the sovereign state of Aramco are opened to public inspection. Still, a close examination of the clues points to the probability that it was the surging profits of the Aramco partners that induced the Saudis to raise their prices. Earlier in the 1970s, it was the Shah of Iran, a "friend" of the United States, who had led the OPEC radicals in raising prices. Late in the decade, it was Saudi Arabia, another "friend" with a moderate image, that took over the role of leader. And Saudi Arabia had eager followers.

The thirteen OPEC oil ministers met in the chic Hotel Tamanaco in Caracas on December 16, 1979. Armed motorcades wound through the clogged streets of the city, five thousand soldiers patrolled the area around the hotel, Venezuelan Air Force helicopters whirred in the sky above it, and reporters anxiously milled around in its bar and lobbies. Inside, the ministers convened in the Presidential Suite occupied by Sheik Yamani for a marathon twelve-hour session. They had lunch sent in—

caviar, shrimps in tarragon sauce, and *pâté de foie gras*—and later Sheik Yamani served English tea and Algerian dates.

Four days later, the cartel had failed to agree on any uniform price or limits on production (production limits guarantee tight supplies and higher prices, a historical lesson learned from Rockefeller's Standard Oil trust). Each cartel member was left to set its own price. Hard-line hawks such as Libya (one of the largest suppliers of crude oil to the United States), Iran, Nigeria, and Algeria raised their prices to as much as $30 a barrel. But Saudi Arabia, though it had instigated the new round of increases, retained its price of $24 despite fierce arguments from the hawks.

Sheik Yamani referred to the fractious proceedings as a "bazaar." One definite result of this longest and most divisive meeting in OPEC history was that while there was no uniform price hike, the average cost of oil was about 100 percent more per barrel than it had been a year before.

What could the United States and other industrialized nations do about it? Pay. There was no other reasonable response under the immediate circumstances. Their governments, and the Seven Sisters, held no leverage over the producing countries, whose only motive for exercising any restraint at all was the desire not to cripple the economies of their customers, who were also the custodians of their investments.

"Insufficient supplies at any given OPEC-administered price level represent a far greater threat to the economic and political stability of importing countries than the price level itself," commented John Lichtblau, executive director of the Petroleum Industry Research Foundation. The "ultimate nightmare," according to the respected Mr. Lichtblau, would be for prices to rise enough to allow the OPEC nations to sell less oil, yet pay all their bills. Production could then be reduced, leading to another price hike, followed by another production decline, setting off still another price spiral. "This self-perpetuating spiral could continue," said Mr. Lichtblau, "until the breaking point—wherever that may be and whatever that may be."

As if to underscore the point, Pemex, the Mexican government oil monopoly, which is not a member of OPEC, raised the price of its crude oil on January 1, 1980, from $24.60 a barrel to $32. A year before, the price of Mexican crude had been $14.10 a barrel. Eighty percent of the oil Mexico exports goes to the United States—about 440,000 barrels a day of total U.S. imports of 8 million barrels a day.

U.S. dependence on foreign oil contains the seeds of another nightmare. That would be a fundamentalist Islamic revolution that overthrew the ruling Saudi family—our moderate "friends" in the area. Since the Shiite Moslem revolution in Iran, with its outbreak of virulent anti-Americanism, the once stable Saudi kingdom—which is governed by Sunni Moslems—has witnessed demonstrations by its small Shiite minority in the oil towns of the Eastern Province along the Persian Gulf in which scores died and hundreds were arrested. In November 1979, the Grand Mosque in Mecca was seized by a large group of armed extremists whose motive was to denounce the Saudi regime for its political and religious corruption. After the Grand Mosque was retaken, Prince Sawaz ibn Abdel Aziz, governor of Mecca, resigned for "health reasons" and Major Fawzi al-Awfi, commander of the Public Security Forces, was ordered to retire.

Such internal dissensions, combined with Soviet intrigue and power in the Middle East and the possibility of another Arab-Israeli war, dramatize how precarious is our future supply of Middle East oil. If the Soviets follow their invasion of Afghanistan with further incursions into the area, either by outright takeover or by setting up a proxy government in one or more of the vulnerable Islamic states, the stage would be set for a confrontation in which oil would become a serious weapon. If oil were denied to the United States under these circumstances or because of American support for Israel, it is doubtful that we could rely on Japan and Western Europe, which are much more dependent than the United States on Middle Eastern oil, to support us.

As the turbulent decade of the 1970s entered its final days, Americans could not be assured of adequate supplies of oil but could be certain that its price would rise in the 1980s. For OPEC and Mexico were sure to raise their prices still further, and the price of domestic oil would finally reach world levels when it was totally decontrolled in 1981.

Again, what can be done about it? Again, the answer is pay. This is a hard answer for a society whose industrial productivity was long boosted by substituting low-cost energy for high-cost labor, that found it economical to construct buildings with minimal insulation because they were once so cheap to heat and cool, and, most of all, that has a sacred tradition of driving large gas-guzzling cars. For the automobile is the main enemy in the energy battle, and will be as long as Americans choose to keep guzzling as much oil as the producers choose to sell. The only realistic alternative, short of gasoline rationing, is for Americans to drive less in smaller cars that burn less gasoline.

There is no technological quick fix that is going to provide a substitute for gasoline over at least the next two decades. Alcohol, liquid fuels from coal, and oil from tar sands or oil shale are a long way from being feasible substitutes from an economic, technical, or environmental point of view. A nonpolluting, inexhaustible source of power like garbage conversion, ocean tides, or solar energy cannot be harnessed to power the automobile. And a rational public transportation system would cost a fortune and take decades to build. So massive conservation and high-performance cars are the only present alternative to remaining an abject beggar for crude oil from unstable societies directly in the path of Soviet influence.

Since people use less gasoline if it costs more—because of a stiff government tax, or free market prices—the depressing and ironic note in the gas-pump blues is that the time has come to rely on price as the mechanism to produce serious conservation. Depressing because that means higher prices. Ironic because that is exactly what the oil companies have been pressing for. It should be remembered, however, that the federal government has subsidized voracious consumption by holding the domestic price of energy below the world price—in contrast to European governments, which let energy prices rise after the first oil shock.

In early 1980, Americans were still getting a bargain compared to motorists in most other countries. True, in Saudi Arabia, gasoline cost 29 cents a gallon, and in Mexico, 65 cents a gallon. But it cost $3.11 in Belgium; $3.02 in Italy; and $3.08 in France. In Great Britain, which exports 28 percent of its North Sea production, gasoline cost $2.34 a gallon and the government had no plans to cut prices.

A return to the precepts of the market economy would encourage vitally needed conservation, even though it would also contribute to inflation and work a hardship on the poor. As for the oil companies, the windfall profits tax on decontrolled prices is designed to prevent them from adding to their already swollen profits in a way that might cause beleaguered citizens to rise up as their forebears did during the Boston Tea Party.

It would be a serious misjudgment to look to the oil companies to find a way to restore plentiful supplies of gasoline at the prices of the good old days. Undeniably, they have the technical expertise to search for new domestic supplies, but they will do so only if they perceive an adequate profit in it, and that means higher prices. Since the time of whale oil, kerosene lamps, and Colonel Drake, the first priority of oil men has been the bottom line. The only difference is that today oil is critical to the national security and economy, and the industry totally controls the supply of this vitally needed product.

Oil may have developed into a business that is too important to be left exclusively to the private oil companies, despite their protests of being hamstrung by excessive government regulation. How far from effective this regulation really is can be surmised from the following assessment in the *New York Times* of January 6, 1980:

> Today, the Energy Department has become one of the most malodorous agencies in the Federal bureaucracy; hundreds of staff positions are vacant, some initiatives stemming from legislation are years from fruition, middle-level bureaucrats are scornful of the political appointees above them and vice versa,

and there is a pervading sense that many years will be required to put the pieces right.

The General Accounting Office, the investigative arm of Congress, in a report to Congress in September 1979 asserted that the oil companies had aggravated the spring and summer gasoline shortages by cutting domestic production of crude oil in the previous fall and winter, though it could provide no evidence that the companies had deliberately created shortages to drive up prices. As for the government agency assigned to regulate the industry and assure the public of adequate supplies of gasoline, the GAO report said:

> Department of Energy actions and pronouncements about the Iranian situation were fragmented and, at times, contradictory. . . . [It] did not provide the Congress and the public with credible, convincing explanations of the status of gasoline, diesel, and home heating oil supplies. . . . The department's lack of adequate energy planning and data has led to inconsistent and conflicting Administration statements and policies on the United States shortfall. It has done little to create a specific plan of action for responding to energy emergencies.

Probably the most bizarre episode involving the Department of Energy took place in December 1979, just after the OPEC meeting in Caracas. It was then that the department chose to auction off 127,465 barrels of oil from the Elk Hills Naval Petroleum Reserve in Kern County, California, once under the control of the Department of the Navy, to private oil companies. More than forty sealed bids were opened just before Christmas. The Department of Energy accepted the high bid from the Phillips Petroleum Company of $41 a barrel for 10,000 barrels a day. Phillips maintained that it needed the crude to meet contractual obligations to a small California refiner. But its purchase of $41-a-barrel oil created chaos in international oil markets, for it came at a time when the Saudis were charging only $24 a barrel, when the Carter administration was urging moderation on OPEC, when the price of decontrolled domestic oil was about $30 a barrel, and when spot market prices were gyrating wildly.

André Giraud, the French minister of industry, called this sale at the world's highest contractual price a "deplorable" mistake because it drove up oil prices. Ray Hnatyshyn, the Canadian minister of energy, pointed to the sale to justify increased prices for Canadian natural gas shipped to the United States. Fred Dutton, a Washington attorney and former State Department official now representing Saudi interests, said that many moderate Saudis were angered because "it made some of them look like fools in the eyes of their countrymen." Edwin Rothschild, director of Energy Action, a Washington consumer activist group that lobbies to hold down oil prices, called acceptance of the bid "an outrage": "It looks like the Government wants to outdo the Arabs in jacking up the price of crude oil."

Domestic producers, who at the time were being exhorted to moderation by the Carter administration, wondered why they should charge less for their decontrolled crude oil than the federal government was willing to take. Frank Cahoon, president of Copano Refining Company in Corpus Christi, Texas, declared: "Accepting the Phillips bid was absolutely stupid because it drove up oil prices overnight, even beyond $41 a barrel."

Secretary of Energy Charles Duncan rather lamely admitted that accepting the $41 bid was a "disadvantageous thing to do" and promised that his department would attempt to stop such a "bad thing" from recurring. In the end, the furor died down only when Phillips withdrew its bid on the (legal) grounds that it could buy the oil for less elsewhere.

The public and the federal government even today do not fully understand the ways in which the major oil companies deal with foreign governments. The United States is the only major Western country that permits private oil companies to manage the procurement and distribution of supplies from abroad—perhaps because of its staunch trust in the free enterprise system. But as exporters exert more control over their oil, consumers are being forced to seek more government-to-government and private supply arrangements at whatever the market will bear because the Seven Sisters can no longer provide the bulk of their needs. Ten years ago, 70 percent of the world's oil passed through the Seven Sisters' hands. Now the figure is less than 50 percent and dwindling.

In buying oil from the OPEC countries and transporting it to the United States, the major companies now function simply as agents for OPEC. The more money OPEC makes, the more the majors make, for the money they earn as agents is a percentage of the total import business. Therefore they have no interest whatever in reducing the dollar volume of the international oil trade. The higher the price of a barrel of oil, the greater will be the dollar volume of that trade and the higher will be the companies' profits. Moreover, no United States government agency really knows what the Seven Sisters, which all market in the United States, do with the crude whose procurement and distribution they have retained control of.

Just so, the figures in the oil industry's record-breaking 1979 profits accepted by the government were computed by the oil companies' accountants using the oil companies' accounting methods.

The IRS recently approved a new method of inventory accounting for all businesses known as LIFO (last in/first out). LIFO has been particularly beneficial for the oil companies because it allows them to price at the most recent market level millions of barrels of oil bought at a lower price and stored in tanks for months or even years. When Saudi Arabia, for example, raised the price of its crude oil from $18 to $24 a barrel, a company that had bought the Saudi crude at the earlier, cheaper price and stored it could—quite legally—overnight charge the $24 price for it. Considering the hundreds of millions of barrels of stored oil, it is not difficult to see what that $6-a-barrel difference did for oil company profits.

Perhaps the time has finally come to make a serious effort to regulate investor-owned oil companies in the public interest, just as investor-owned public utilities are regulated. The theory behind utility regulation was that the public was at the mercy of private companies for essential services such as electricity, and it was impossible, for practical purposes, to ensure free competition in those services. Today a handful of majors handle the importation of foreign oil, which is certainly essential to the whole society. But no amount of regulation will erase a central fact of life: The root cause of our energy problem is that Americans consume far too much gasoline. If we, and our elected representatives, can't face that fact and resolve at last to do something about it, we will have to take the economic and political consequences. And they won't be pleasant. In 1979, America imported $90 billion worth of petroleum, in effect putting its neck in an economic noose held by foreigners. Will Rogers once said that America was the first nation to drive to the poorhouse in an automobile. Unless those automobiles and the Americans who drive them stop burning up so much gasoline, Will Rogers may turn out to have been a prophet.

The oil companies, to be sure, will not end up in the poorhouse. But it is fruitless to berate them for a situation that our inquiry shows was caused by a complex series of national and global events. In separate, detailed studies made by the Justice and Energy departments and released in July 1980, it was concluded that the industry's management of crude oil and gasoline inventories played no significant role in the 1979 shortfall that had motorists at one another's throats.

What happened is easy enough to follow. When Iranian production faltered and then collapsed altogether, this markedly reduced total world oil supplies and caused a global scramble for what was available. Although Iranian crude represented only a small portion of U.S. imports, its cutoff was enough to cause a shortfall. The latter was then mismanaged by the Energy Department's complex allocation scheme, which, for various reasons, failed to match available supplies accurately with demand.

Only a year after lines began at service stations, there was plenty of gasoline and world prices were actually sagging a little. To those who savor conspiracy theories, it was suspicious that there was more than enough gasoline now that the oil companies had gotten higher prices and were reaping record profits. But again, it is easy enough to follow what happened. Saudi Arabia's increased production made up for the Iranian shortfall. American and European industry, caught in the doldrums of a recession, burned less fuel oil. Motorists finally began to cut back on their use of gasoline. In the first three months of 1980, Exxon's sales of gasoline (but not its profits) were down 12–13 percent from a year earlier. There was, in fact, a world surplus of crude oil and the storage tanks of the oil companies were bulging.

Chapter Eight

Who's Got the Oil, and for How Long?

The United States will not run out of oil next year, or the year after, but the production decline that began in 1970 will never be reversed. The mainland United States has been thoroughly explored and we have passed the point of no return in witnessing dramatic new discoveries of giant fields. So the question is: Who's got the oil now, and how long will they have it?

To answer this question, we must enter a world where things are not always what they seem, the world of petroleum statistics. Here we encounter such terms as "proven" and "probable" recoverable reserves, "reserves in place," and "resources." One might think that all these terms mean more or less the same thing, but not so. To oil men, each of them has a different meaning.

"Proven" reserves are the only really dependable guideposts. They are official estimates (revised periodically to reflect new data) of untouched oil and gas deposits that geological and engineering data show are reasonably certain to be recovered or extracted, given the prevailing technology and price. Experts maintain that these estimates are accurate to within 10 percent.

Now we come to estimates in which less faith should be placed, those of "probable" reserves. They are projections of how much oil can be pumped from already discovered fields that have not yet been completely explored and developed—again, providing the price is right. Exxon, for example, is not going to exploit "probable" reserves at a cost of $40 a barrel if it can sell the oil for only $30 a barrel, no matter how extensive the reserves might be.

"Reserves in place" means all the oil included in the "probable" category, but with the qualification that it may not be recoverable with current technology and at current prices. This oil will simply be left "in place" unless the technology to recover it at a profit is developed.

Lastly, there are the estimates of what are called "resources." This is oil that has not yet been found, or has been found in such remote places that it can't be transported and sold to consumers, such as in the High Arctic. Estimates of "resources" are really nothing more than educated guesses.

It may comfort the reader to know that, apart from the "proven" recoverable reserves, government agencies, private geologists, academics, and oil companies don't know a great deal more than the rest of us about how much oil is left on the planet and disagree widely about their own guesstimates.

The Vanishing American Birthright

The United States was once the world's biggest oil producer, which made possible the growth of its economy, the automobile age, the interstate highway system, and suburbia. American output peaked in 1970, at a time when demand was rising rapidly. Although government policies, overconsumption by consumers, and the drive by oil companies for profits all contributed to the shortage, its prime cause is based on elementary principles of geology. Accustomed to regarding cheap and abundant oil supplies almost as a birthright, Americans have disregarded the law that nothing last forever in the world of finite resources. A few petroleum geologists and economists had for years been calling attention to the approaching exhaustion of U.S. reserves, but their analyses, published in academic journals or reports of congressional hearings, were ignored or soon forgotten. As the eminent petroleum geologist M. King Hubbert put it in 1974: "The initial supply is finite; the rate of renewal is negligible; and the occurrences [of discovery] are limited to those areas of the earth where the basement rocks are covered by thick sedimentary deposits. . . . A child born in the 1930's, if he lives a normal life expectancy, will see the United States consume most of its oil during his lifetime."

In 1969, Texas produced 35 percent of the total U.S. output of 14.4 million barrels a day. To maintain that 35 percent share in the future, it would be necessary for Texas to produce 5 million barrels a day. Judge Jim C. Langdon, chairman of the Texas Railroad Commission, which regulates the oil industry in that state, was asked about this possibility at a hearing before the Senate Antitrust Subcommittee. The judge replied: "We would not even pretend to try to produce that much oil simply because under existing conditions *we could not produce this much oil*. . . . We are not going to contest one iota the statement that we are going to be more dependent in 1975 and 1980 on foreign crude than we are today."

During the same 1969 Senate hearing, M. A. Wright, vice president of Exxon took a much more optimistic view of future domestic discoveries. He was questioned by a dubious Senator Hart of Michigan:

> SENATOR HART: Mr. Wright, let me interrupt you. You say that this future reserve—which by 1985 must be relied on for 85 percent of our needs—represents oil which *will* be found. That is the way you put it.
> MR. WRIGHT: That is true.
> SENATOR HART: How do the bookmakers quote the odds? That is an awful lot of oil that somebody says will be found . . . are you not playing sort of Russian roulette with national security?
> MR. WRIGHT: I do not think so, Senator.

If he was not playing Russian roulette, the Exxon vice president's crystal ball was definitely clouded, for a decade after those 1969 hearings, substantial new proven reserves had *not* been found except on the North Slope of Alaska—and that bonanza has proved to be a mixed blessing. The prospect facing the United States at the start of the 1980s is one of falling production, declining proven domestic reserves, increasing dependence on unstable imports unless consumption is drastically reduced, and potential disaster to the environment when offshore production is fully exploited.

No one can state with certainty when America's oil will run out in the next century. But informed projections about what is likely to happen through the end of this century have been made, one of them by Maurice ("Butch") Granville, chairman of the board of the American Petroleum Institute from January 1976 through December 1977 and chairman of the board of Texaco, the nation's third largest oil company and one of the Seven Sisters, from 1971 to 1980:

> We are drilling more and finding less. Until 1950, we had a high oil-finding rate per exploratory foot drilled. More recently, the oil-finding rate has been only a fraction of the rate during this earlier time. Most of the so-called "easy oil" and gas in the shallow accumulations has been found. We have to drill more feet for each barrel of oil and each cubic foot of natural gas we discover.
>
> There's an old saying in our business: When you start drilling for oil you may be a thousand feet from a million dollars or a million feet from a thousand dollars. We don't see enough of those thousand-foot discoveries these days.

Mr. Granville sees 1985 to 1990 as the watershed period in which worldwide availability of petroleum will decline, with consequent shortages for consumers. "There is now much speculation as to when Free World and American production will peak out, and there will continue to be until that inevitable moment actually arrives," he says. "There are those who still believe that somewhere, somehow, a new Middle East will be found, or that deepwater drilling will uncover vast new reserves." But the former Texaco chairman is skeptical that it will occur in time to prevent Free World petroleum supplies from topping out in the mid-1980s.

Is Butch Granville being realistic, or pessimistic? Everyone knows that huge strikes have been made in Alaska, and Alaska is, after all, part of the United States. Can't Alaska make up for the production decline in the lower forty-eight states?

Cold News from the North

The 10-billion-barrel oil pool at Prudhoe Bay on Alaska's North Slope, discovered in 1968, is the nation's largest oil field by sevenfold. By early 1980, 1.5 million barrels a day were flowing 800 miles south through the Trans-Alaska Pipeline System (TAPS) to the port of Valdez, where tankers loaded and transported most of it to refineries in California and the state of Washington. There's so much of it that these refineries can't process it all, so the surplus is shipped through the Panama Canal to Gulf of Mexico and East Coast ports.

The U.S. Geological Survey has come up with a mean estimate of 30 billion barrels of Alaskan oil yet to be discovered, more than one-third of the nation's total. The USGS also estimates that Alaska contains 76 trillion cubic feet of undiscovered natural gas, or 16 percent of the U.S. total. Hopes are high for future big discoveries under the American sector of the Beaufort Sea, 275 miles above the Arctic Circle off Alaska's north coast.

Alaska has been a bonanza for the few oil companies prescient enough to have bought Prudhoe Bay leases and for at least some U.S. citizens. The Los Angeles–based Atlantic Richfield Company, which, along with Exxon, is a major owner at Prudhoe Bay, saw its earnings in the first quarter of 1980 soar 75.9 percent over the year-earlier first quarter and its revenues climb from $3.52 billion to $5.43 billion.

Citizens' rewards have been more modest, but Alaskans are undoubtedly the only Americans who have reason to applaud when OPEC members, dutifully followed by the oil companies, raise prices. That's because the state's oil royalties grow with each price hike. In 1980, some 270,000 Alaskans had the novel experience of receiving checks averaging $1,000 each at income tax filing time from the state's growing oil revenue fund.

All this is good news for Atlantic Richfield and the citizens of Alaska. But is it good news for the rest of us in the sense that we can expect Alaska to provide a solution to the decline in U.S. production? The melancholy answer is that this bonanza can only arrest the decline for a few years.

The Central Intelligence Agency made a blunt assessment in its research study *The World Oil Market in the Years Ahead,* circulated in August 1979:

> Prudhoe Bay—the largest field by far ever found in the United States—pales by comparison with current levels of oil consumption. If it could be produced at will, Prudhoe Bay would satisfy world consumption for five months. It would satisfy US oil consumption for only 16 months. . . . After 1985, output from the Prudhoe Bay field will also begin to decline rapidly, according to reservoir studies done for the Alaska State Legislature.

Well, then, what about the natural gas associated with the Prudhoe Bay oil deposits? There are at least 27 trillion cubic feet of known recoverable gas—more than a year's national supply and 10 percent of the nation's gas reserves—and authoritative estimates suggest a comparable additional amount can probably be extracted. These vast deposits are not, of course, of any use so far as gasoline consumption is concerned, but they can be substituted for oil in industry and for home heating. Unfortunately, they remain where they were found—in the ground.

The projected construction of a 4,800-mile pipeline system to transmit the gas from the North Slope all the way across Canada to the lower forty-eight states—prospectively the biggest civil engineering project in history—has been stalled by a variety of problems, chiefly financial. The gas industry, which is a service business without the huge assets of the oil industry, has no way of financing what may well be a $15 billion undertaking from its own resources. Federal financing, which would shift the financial burden from gas consumers to taxpayers, may be the only way of revitalizing the project—but federal financing is now prohibited by law. Here again, we come up against the disheartening fact that as the country runs out of oil and gas, private industry cannot be induced to produce new supplies and bring them to market unless it is assured of a reasonable profit for doing so.

So far, some oil companies have profited quite nicely from Prudhoe Bay, such as Exxon and Atlantic Richfield, the major owners, and Mobil, Phillips, Union, and Amerada Hess, which have smaller interests. Even the British government makes money at Prudhoe, since it owns 15 percent of its production. This bizarre circumstance arose because the British government owns 51 percent of British Petroleum, Ltd., which in turn owns 52 percent of Standard Oil of Ohio (Sohio), which works the lease as an affiliate of BP. But many other oil companies have not found Alaska to be a bonanza. Plagued by expensive exploratory failures (Marathon drilled one dry hole at a cost of $38.5 million for the lease alone), high transportation and operating costs, meager profits, state taxes and environmental disputes, they have simply given up, as did earlier waves of prospectors seeking gold, furs, and whales. According to the Alaska Oil & Gas Association, twenty-one energy companies have backed away from Alaska since 1971. As Roger Bexon, president of Sohio Natural Resources Company, put it: "At the time Prudhoe Bay was discovered, expectations ran riot. People thought there might be a dozen of these things lying around and all sorts of smaller ones. People have now come down to earth."

On this note of deflated expectations about Alaska, let us return to the lower forty-eight states.

The Lower Forty-eight

Have you ever met a well-log analyst? If you have, you know that he's a crucial person in petroleum exploration. For when a wireline, or armored cable, is lowered down a borehole to a depth of as much as 3 miles, it is up to him to interpret the maze of electronic signals transmitted by the cable and tell the driller if they all add up to enough hydrocarbons to indicate a producing well. The arcane trade of the well-log analyst demands a thorough training in engineering, geology, and physics.

Then there are the jug hustlers, thumper truck drivers, and doghouse observers—all colorful names for those engaged in doodlebugging, the seismic mapping that employs a new generation of instruments and field techniques to discover oil in the lower forty-eight states, where over 2 million wells have already been sunk. The jug hustlers, or jerbs, lay black cables as much as 4 miles long over a survey site. Attached to the cables are fist-sized geophones, called jugs, which are driven into the earth at intervals. Then the thumper truck drivers arrive. Instead of containing the dynamite that was used to blow apart the earth in the old days, their trucks house vibrators, or thumpers, that press a steel plate to the ground, vibrate it, and transmit seismic waves from deep below the surface to the doghouse, or recording van. In the doghouse, an observer directs the operation and interprets the vibrations and sound waves sent up to rows of computers by the jugs and thumpers. Naturally, the doghouse observer hopes that all of these reflected electronic impulses will suggest that oil is trapped somewhere down there in the substrata.

As higher petroleum prices alter the economic equations, these seismic crews are going to areas that once were considered too marginal to bother with, probing into deeper and more complex geological formations than ever before. The pace of oil and gas drilling has quickened markedly in regions outside the old "mature" provinces. More than twenty-five hundred drilling rigs are now in operation on and offshore.

The most promising of the new regions is the Overthrust Belt, a geological formation that runs from Canada down through Idaho, Wyoming, Utah, Colorado, and Arizona, where, because the deposits are deeper, it costs five times as much to drill a well as it does in Texas and Oklahoma.

Oil fever has even gripped areas of Appalachia where no oil has ever been produced. The so-called Eastern Overthrust Belt, a geologic formation with similarities to the one in the Rockies, runs 1,000 miles from New England through West Virginia, Kentucky, and Tennessee to end in Alabama. Although the Eastern Overthrust Belt has plenty of coal, until recently the idea that oil might also exist there in commercial quantities was considered a geologic impossibility. Now 10 million acres have been leased by major companies like Exxon and Gulf and wildcatters like George Simms, a man in his late forties who abandoned a real estate development business in Providence, Rhode Island, to search for oil in Deer Lodge, Tennessee, in 1971.

In 1979, Mr. Simms struck oil in commercial quantities on the 200-acre farm of Sergei Luchin, setting off a flurry of exploration in the strip-mined hills along the Tennessee-Kentucky border. Mr. Simms's motivation, like that of other wildcatters, was basic. "In 1971, when oil was going for $2.80 a barrel, you had to have an enormous oil well to entice the investment capital needed to back a drilling operation," he said. "Now, with oil at $26.50 a barrel, well, that kind of price makes it more interesting."

Wallace DeWitt, a scientist with the U.S. Geological Survey who monitors activities in the Eastern Overthrust Belt, strikes a note of caution about what the area might yield. "Everyone is trying to keep his competitors in the dark just like in a high-stakes poker game," he has said, adding: "You really couldn't call this a frontier area yet, but it certainly has potential."

Potential also exists in offshore areas like the Baltimore Canyon in the Atlantic Ocean, the Gulf of Mexico near southwestern Florida, the California coast in the Santa Barbara channel, and the coast of Maine. We will assess this potential in Chapter 9.

Surely, you might think, this salubrious blend of virgin territory, computerized analysis of seismic data, advanced drilling techniques, and huge investment made possible by skyrocketing oil and gas prices has got to pay off in restoring the good old days when gasoline was 30 or 40 cents a gallon, with a set of dinnerware thrown in, and we didn't have to depend on Arab sheiks and other untrustworthy foreigners. Unfortunately, virtually no one in the oil business really believes that this new activity will reverse the long-term drop in domestic oil production or arrest the steady decline of domestic probable recoverable reserves. All it can reasonably be expected to do is to prevent domestic production from falling even more sharply in the years ahead—and, of course, increase oil company profits, some of which will be skimmed off by the windfall profits tax. As Ronald McCormick, manager of exploration research at Conoco in Ponco City, Oklahoma, says: "We don't expect to find any panaceas. All we do is sharpen our guesses and decrease the risks when we drill a hole."

If there are no panaceas, are there trustworthy speculations about *when* the United States is going to run out of the resource upon which it has traditionally relied as the prime mover of its economic system? An exact date cannot be predicted, of course, and the estimators always temper their calculations with qualifications concerning such vagaries as demand growth and energy/gross national product ratios. But leaving aside such economic esoterica and divorcing ourselves from political rhetoric, wishful thinking, and oil company public affairs ads, we can say that (1) the United States has been losing oil far faster than drilling crews have been finding it; (2) production will peak between 1985 and 1990; and (3) the wells will inevitably run dry, probably toward the middle of the next century.

From the oil companies' side, Exxon has projected U.S. oil supply at only 8.5 million barrels a day in 1985, and as little as 7 million barrels a day in 1990.

The Central Intelligence Agency, in its 1979 research study, forecast a decline in U.S. production and noted that the pace of onshore and offshore discoveries has been discouraging: "Despite a 65-percent increase in drilling in the lower 48 states since 1973, new discoveries have been small and the finding rate has dropped precipitously."

Owen Phillips, Decker Professor in Science and Engineering at the Johns Hopkins University, has written:

> In fifty years we have used over half of all [the oil] there ever was, and at the rate we're going, we'll finish off the rest very quickly. . . . Even allowing for other large finds not yet contem-

> plated [such as Prudhoe Bay]—and there cannot be too many more of these—we might add a few more years, maybe ten at the outside, but the conclusion is inescapable. The United States is going to run out of oil very soon.

Robert Stobaugh summed up the situation succinctly in *Energy Future—Report of the Energy Project at the Harvard Business School,* which he edited with Daniel Yergin:

> Americans should not delude themselves into thinking that there is some huge hidden reservoir of domestic oil that will free them from the heavy cost of imported oil. Of course, reasonable measures should be taken to keep domestic oil production from declining further. But the handwriting is clear. To the extent that any solution at all exists to the problem posed by the peaking of U.S. oil production and the growth of imports, it will be found in energy sources other than oil.

So the consensus of informed opinion is that we will have to continue importing oil. The question then becomes: How much oil is there in the rest of the world, and where are we likely to get it on the most advantageous and reliable terms? In view of the political tinderbox the Middle East has become, should we look to our friendly neighbors to the north and south and a new entrant into the world oil game a bit farther away with whom we maintain friendly relations, China?

The Mexican Connection

Canada cannot be of any significant help in supplying the United States with oil because proven reserves peaked there in 1969 and production has declined since 1973. In 1975, Canada became a net oil importer and discoveries in its remote frontier regions like the Arctic Islands have been mostly natural gas.

The situation is different south of the border. There has always been oil in Mexico. International companies, mostly American, once exploited the great oil fields of Vera Cruz, along the western coast of the Gulf of Mexico, so rapidly that the original deposits were destroyed without hope of recovering what was left. Mexico responded on March 18, 1938, by expropriating the foreign holdings and forming Petróleos Mexicanos (Pemex), the first national oil company. The state-owned Pemex is the only oil company permitted to operate in Mexico and is basically self-sufficient from the wellhead to the gas pump, although it does buy about 70 percent of its equipment from U.S. companies.

The nationalization didn't make much difference to American supplies in 1938, given the extent of U.S. reserves, though it was a blow to the oil companies and their stockholders. Now, though, Mexico is sitting on a pool of oil and gas second in size only to the reserves of Saudi Arabia.

The emergence of Mexico as an oil superpower has been recent and dramatic. Pemex made its first big discovery in Villahermosa in May 1972, and followed with others in the southeastern states of Tabasco, Chiapas, and Campeche in 1977. Two years later, Mexico's proven oil reserves had jumped from 6 billion to 40 billion barrels and were rising fast.

The man who runs Mexico's booming oil operations is Jorge Díaz Serrano, a childhood friend of Mexican President José López Portillo, who appointed him director general of Pemex in 1976. A soft-spoken engineer who worked for the national irrigation commission for two years, Mr. Díaz Serrano won a scholarship to the University of Maryland from the U.S. Department of Commerce. He then joined the Chicago Pneumatic Tool Company for a year and a half before returning to his homeland to join Fairbanks Morse. Later he formed a number of private ventures, some of which supplied Pemex.

The genial Mr. Díaz Serrano, who conquered alcoholism in the 1960s, always had faith that there was still a lot of oil to be found in Mexico and that large-scale production would offer a way out of a miserable socio-economic situation in which nearly half the Mexican people are unemployed or underemployed and abysmally poor. He has been proved right about the reserves, but what does this mean for oil-thirsty America?

In November 1979, Mr. Díaz Serrano attended a meeting of the American Petroleum Institute in Chicago and announced to a related group of oil explorers called the Nomads the discovery of the Chicontepec field. He described it as "one of the bigger hydrocarbon accumulations in the Western Hemisphere" and estimated it could contain 100 billion barrels of oil.

If that estimate is correct, when added to Mexico's proven and probable reserves, it would give the country a total of some 300 billion barrels that could one day actually materialize and be marketed. In 1980, 80 percent of Mexico's oil exports went to the United States, about 440,000 barrels a day of total U.S. imports of 8 million barrels a day. By 1982, Mexico expects to raise its daily production to at least 2.25 million barrels a day, but has placed a ceiling of 1.1 million barrels a day on total oil exports through that year.

Unfortunately, Mexico's immense reserves do not mean that long-term U.S. import needs will be filled at reasonable prices. Although Mexico has declined membership in OPEC, it does follow that organization's pricing policies. In January 1979, the price for Mexican crude was $14.10 a barrel. It went to $17.19 in March, $22.60 in July, $24.60 in October, $32 a barrel in January 1980, and $38.50 a barrel in December 1980.

Even American willingness to pay those prices will not guarantee us all or most of the Mexican output, for other countries are eager buyers, including Spain, France, Israel, and Brazil. In May 1980, Prime Minister Masayoshi Ohira of Japan proposed a deal to the Mexican government (unsuccessfully, as it turned out) whereby Japan would secure a significant increase in the meager 22,000 barrels a day Mexico allows it to import in exchange for greater Japanese investment and technical assistance.

It is dangerous to assume that the Mexicans will be able or willing to sell as much of their patrimony, their *petróleo,* as the United States wants to buy, for other reasons. The Mexicans have neither the capacity to produce all the oil they've discovered nor the infrastructure—the transportation, ports, materials, and support services—to send it north. Then there are political problems. Mr. Díaz Serrano has had to fend off the charge by

Mexican leftists that he wants to sell out Mexico's patrimony to the United States. There have been disputes between the American and Mexican governments about how much Mexican natural gas the United States should buy, and at what price. Lastly, there is a growing consensus among Mexican officials that output should be restrained to levels commensurate with Mexico's ability to absorb the resulting massive revenues. "The capacity for monetary digestion is like the human body," President López Portillo observed in 1979. "You can't eat more than you digest or else you become ill."

Foreign Minister Jorge Castenada has commented that Mexico should avoid sending too much oil to the United States, "not so much because of the risk that an excessive share of sales to the U.S. would increase our dependence but because of the greater danger that the U.S. would become excessively dependent on Mexican oil." That the Colossus of the North is in danger of becoming economically dependent on Mexico is a startling thought for most Americans.

The sensational Mexican discoveries, therefore, are only one more welcome development that will ease but not eliminate the supply shortages predicted for the United States in the latter part of this century. Mexican production will postpone the projected doomsday of declining world oil supplies into the next century, but it will definitely not cancel it. In 1979, John Swearingen, then chairman of both the American Petroleum Institute and Standard Oil of Indiana, observed that even if Mexico produced 5 million barrels a day, that would represent the equivalent of only about a three-year growth in world oil consumption.

It is interesting to reflect on where the United States would be today if the American oil companies had known in 1938 how to control the flow from wells to maximize the amount of oil ultimately recoverable from the old fields of Vera Cruz. If they had not opted for torrential, and destructive, production, they might not have been ejected from Mexico and perhaps would still own all that oil—or at least have a voice in what happens to it, instead of standing, Stetson in hand, in line with all the other customers for what Pemex chooses to sell at its own price.

The China Syndrome

When they were still allies, China got most of its oil, and oil technology, from the Soviet Union. Until the reopening of China to the West in the early 1970s, American oil men faced a Great Wall of ignorance about that country's oil potential. Now, as part of the ambitious Four Modernizations program, government ministers see development of China's oil and gas reserves as a way to gain the capital they need for industrial development.

U.S. geologists, oil companies, and technology are no longer considered the running dogs of imperialism but are welcome in China. Though much of the country remains unexplored, guesstimates place current proven and probable reserves at around 20 billion barrels.

In June 1979, China's highest economic official, Kang Shien, a deputy prime minister in charge of energy and economic planning, visited Washington seeking cooperative ventures with American oil companies, especially for geophysical prospecting offshore. In November, Coastal States Gas Corporation bought some $50 million worth of low-sulfur Chinese crude for its California refinery. Then, in mid-March 1980, Chem-Oil, a small, privately held New York–based trading company, became the first Western company to buy refined petroleum products from China. However, the nearly 13.5 million barrels of gasoline and diesel fuel purchased for $50–$60 million was not destined for the United States, according to Chem-Oil. It would most likely be sold in Africa or East Asia, where the economics were more favorable.

Later that March, China allowed a group of U.S. and European oil men to inspect its newest oil field, the Renqiu field in Hebei Province, about 94 miles south of Peking. This field had first been developed in 1975 and was now producing about 20,000 barrels a day. In April, the official New China News Agency reported the discovery of six major new oil deposits expected to produce a total of 21 million barrels of oil a year around the Daqing field.

Does all this activity mean that China might become another Saudi Arabia or Mexico and a major supplier to the United States? Whatever happens, it is certain that Houston will play a major role. The production of oil and gas has been selected as China's "model" enterprise in its struggle to become a modern industrial power. Houston, the acknowledged world center of petroleum technology, has the equipment and expertise that China needs and wants and can no longer get from the Soviet Union.

Houston companies have already done nearly $250 million in business in China since 1973, and the action picked up when the Chinese government established two new consulates in the Texas city. This is the same Houston whose inhabitants have so often expressed right-wing, militantly anticommunist views. The Peking-Houston connection is but another illustration of how world politics and the oil business are intertwined and how political convictions can be bent—on both sides—to meet economic priorities. Says Texas Governor William P. Clements: "They're going to have to buy from us the kinds of equipment that will enable them to develop those oil and gas reserves. So that's good for business here in Texas. I don't see anything but pluses in it, frankly." Mr. Clements is a conservative and the first Republican governor of Texas. Before he became governor, he was the president of SEDCO, the world's largest drilling contractor.

Despite the ideological accommodation, China's reserves may never be developed in vast commercial quantities, and its oil may never find its way here in sizable volumes. The CIA believes it unlikely that China will become a major supplier of crude oil in the 1980s. According to its 1979 research project:

> Output of 2 million b/d in 1979 puts China among the important world producers—comparable to Libya and Nigeria. But domestic demand is rising rapidly, and China already consumes some 90 percent of its own production. . . . We believe that the amount of exportable oil might level off at about 300,000 b/d in 1982 or so. To increase exports beyond 1982, China would need

> considerable luck in locating large and easily exploitable reserves or would have to enforce stringent economies in domestic oil consumption. One problem is that large-scale offshore production can probably not begin before 1985.

So it would appear that China will not be a major factor in closing the gap between domestic production and U.S. requirements, or at least not until the end of this century. That role, for better or worse, will continue to be played by the Persian Gulf countries, which hold about 60 percent of the world's 650 billion barrels of proven recoverable reserves.

Inexhaustible Suppliers?

Saudi Arabia, with a population of some eight million (compared to China's more than one billion), is the world's undisputed monarch of world oil. It has proven reserves of 150 billion barrels—one-quarter of this planet's total. The Soviet Union is second with 75 billion barrels, followed by four Moslem Persian Gulf nations: Kuwait, 67 billion; Iran, 62 billion; Iraq, 34.5 billion; Abu Dhabi, 31 billion. The United States ranks seventh with 29.5 billion.*

The crucial importance of Saudi Arabia can be appreciated when it is realized that just one mammoth field there—the Ghawar—accounts for 9 percent of world crude output. In 1980, the United States will get 18 percent of its imported oil from Saudi Arabia. The Saudis supply one-third of the oil imported by Japan, which has none of its own. Obviously, the economies of the world's leading industrialized nations would be shattered by a cutoff in Saudi Arabian production, or a severe disruption, as happened in Iran.

Fortunately, Saudi Arabia is a conservative, strongly anticommunist kingdom that purchases arms from the United States,

* All 1979 figures from *Oil & Gas Journal* and *World Oil.*

has led regional efforts to stabilize the Persian Gulf area, and has counseled moderation in OPEC price hikes. The Saudis even increased their production to offset the Iranian shortfall, although they had no need for the extra income. But what are the prospects for the long-term productivity of the huge Saudi Arabian fields upon which the United States, Japan, and Western Europe are so dependent?

Not reassuring. Under subpoena, two members of the Aramco consortium, Exxon and Standard Oil of California, submitted documents to the Senate Foreign Relations Committee in March 1979 that raised serious doubts about long-term Saudi productivity. (Standard Oil of California's management might well have thought back wistfully to the era when their company, which made the first strikes in Saudi Arabia, was obliged to take in Texaco as a partner to help market the oil because it was so cheap and plentiful.) The documents reported that production from the Saudi oil fields had been hampered by technical difficulties and that if those fields were required to produce 14 to 16 million barrels daily—which some experts predict will be needed to avert shortages in the mid-1980s—they could sustain that level for less than ten years. Citing a study compiled for the Saudi government by a British consulting company, the documents showed that the oil fields, if producing at 8.5 million barrels daily, would begin to diminish around the year 2000. If producing at 12 million barrels daily, the Saudi fields would begin to become depleted within fifteen years. Aramco has not made any significant findings of new reserves in Saudi Arabia since 1970, nor is it likely to do so.

More pessimistic still is the CIA report: "Even the vast reserves of Saudi Arabia, where 11 of the world's 33 supergiant fields are concentrated, would supply world consumption for only about seven years at current consumption rates."

There is no guarantee, of course, that Saudi Arabia will supply the United States and the rest of the industrialized world at current consumption rates forever. In one more disquieting development, the government of Saudi Arabia made the final payment for the refining and producing facilities of Aramco in September 1980. Since 1974, it had owned 60 percent of

Aramco; now it owns 100 percent. Aramco, which once played a pivotal role in world trade, has been demoted to little more than a service organization working for a fee to produce the Saudis' oil and with the right to buy much of it.

According to Aramco's 1980 "Review of Operations," which is the closest thing to an annual report the company issues, Aramco produced a record 3.38 billion barrels of Saudi oil in 1979. But like the other Persian Gulf nations, Saudi Arabia will undoubtedly try to conserve its oil supplies for coming generations by imposing an artificial ceiling on future production. It makes economic sense for them to leave some of their oil in the ground for the long haul instead of converting it to dollars that keep losing value but must be invested.

Saudi Arabia, Kuwait, Abu Dhabi, Qatar, Bahrain, and the United Arab Emirates are about the only OPEC members that sell more oil than they need to in order to balance their bank accounts. But they are running out of satisfactory long-term investment outlets in the West for their surplus petrodollars that provide both a present hedge against inflation and a future secure rate of return for when the oil fields finally run dry.

In 1979, for instance, Saudi Arabia exported 3 million barrels a day more than it needed to, earning some $15 billion in cash. Kuwait, with an estimated population of 1,275,000, earned $5 billion more from oil exports than it spent in 1979. It negotiated a tougher generation of sales contracts in April 1980 and cut its oil production by more than a quarter, reducing its allotments to Gulf, Royal Dutch/Shell, and British Petroleum, the three majors that contracted for two-thirds of Kuwait's average daily production, by nearly a million barrels a day. "We were very glad to have it. Nowadays you take what you can get," said a BP spokesman.

Sheik Ali Khalifa al-Sabah, Kuwait's oil minister, allowed that he would like to sell the oil for more than mere money. "Customers will not just come in, pick up their barrels and pay the OPEC price. They must make interesting [offers] in regard to other things," he observed, mentioning making refineries available for purchase, developing new exploration projects, and other "added advantages for Kuwait somehow."

The autocratic Persian Gulf sheiks appear to have learned a lesson from the Iranian revolution: High oil production and a huge influx of money do not guarantee political stability—on the contrary, there is a danger in becoming a dumping ground for unwanted Western goods and ideas. Today the Iranian revolution is a powerful influence in the area and the conservative rulers fear it will ignite unrest among their own citizenry or their large and underprivileged expatriate populations. The latter include many Palestinian refugees and members of Iran's predominant Shiite Moslem sect. The danger focuses on Saudi Arabia. With its small population, high income level, and generous welfare spending, Saudi Arabia does not have the same social problems as poorer and more populous Iran, but it does have problems. The country has no free press, no parliament, and its streets have no names. The government is the royal family. It numbers about three thousand adult males, about sixty of whom form an inner circle that rules the country. "Saudi Arabia is perhaps best described as the only family-owned business recognized at the United Nations," according to Peter Iseman, an Arabist who has written on Saudi affairs.

The House of Saud belongs to the puritanical Sunni sect of Islam. But in the eastern provinces, where the oil facilities are located, the inhabitants are Shiite Moslems, like those in power in Iran. The Shiites believe that they are treated as second-class citizens by the Sunnis.

The puritanism of the ruling House of Saud does not extend to Western concepts of corruption, including family aggrandizement, conflicts of interest, and bribery demands that would have given Boss Tweed pause. Most princes function as legitimate business agents, but some take multimillion-dollar commissions from Western companies eager to do business in Saudi Arabia and earn multibillion-dollar profits in land speculation. Until now, profiteering has been taken for granted in Saudi Arabia. " 'Corruption' often exists in the eye of the beholder," according to Mr. Iseman, "and Americans and Saudis view the problem with very different definitions and cultural perspectives. Saudis, for example, are greatly concerned with the ideas of Islamic justice and Bedouin sharing, but our legal concept of conflict is quite alien to their personalized and tribal society."

This insight is undoubtedly accurate and makes unlikely the kind of popular uprising that ousted the Shah. However, when the Grand Mosque in Mecca was seized on November 21, 1979, the first day of the fifteenth century of the Moslem calendar, it took the Saudi government two weeks to dislodge the political and religious terrorists whose complaints about the House of Saud included corruption. It is doubtful that further trouble can be avoided as traditional values are challenged by modernization programs (the telephones now work), unrest grows among the Shiites of the eastern oil province, and a Western-educated elite and a professional independent military capable of a coup grow in numbers and maturity. In January 1980, James Akins, a former U.S. ambassador to Saudi Arabia, told a group of financial analysts that "without dramatic internal reforms, the country faces serious problems, as the feelings about corruption are similar to developing feelings in Iran in 1976–77."

One of the princes whose business activities offer a paradigm for review is Mohammed bin Fahd, son of Crown Prince Fahd. The latter is a half-brother to the ailing King Khalid, and as first deputy prime minister has been the day-to-day ruler of the kingdom since the assassination of King Faisal in 1975.

Prince Mohammed is in his mid-thirties and is a graduate of the University of California at Santa Barbara. He operates a wide range of businesses that include the ownership and chartering of oil tankers. He helped Phillips, the Dutch electronics concern, to obtain a telecommunications contract with the Saudi government as the leader of a consortium. Though the original $7 billion proposal called for the payment of more than $100 million to Prince Mohammed, this proved too much for even the Saudis, so a final competitive bid was accepted at half the original proposed price, with no record of the fee the prince received.

Prince Mohammed also joined forces with Bechtel Inc., a San Francisco–based construction company, by becoming a partner in Arabia Bechtel in 1977. The next year, Arabia Bechtel was awarded a $3.5 billion contract to build a new airport near

Riyadh, the capital of Saudi Arabia. It should more than satisfy the transportation needs of a desert kingdom of eight million people and its visitors, since it will be the largest airport anywhere in the world.

None of this would be anybody's business but the Saudis' if it weren't for their oil. Although the State Department is troubled by Saudi corruption, U.S. officials are loath to urge the royal family to make any reforms for fear of irritating Saudi bureaucrats, who are viewed as a key to stability in the region. As for the oil companies, no one expects them to jeopardize their principal source of foreign crude by calling for reforms.

The Aramco partners still have a very good thing going for them in Saudi Arabia. Exxon, Mobil, Texaco, and Standard Oil of California import almost all the Saudi oil that enters the United States, where collectively they have about one-quarter of the petroleum market. When the Saudi Petroleum Ministry raises its price, as it did in mid-April 1980 from $26 to $28 a barrel, the companies raise their gasoline prices 1 to 4 cents a gallon. After all, the Saudi oil still costs less than other OPEC oil of comparable quality.

Saudi Arabia is by far the largest dollar holder in OPEC. Its relative price moderation has stemmed from its rulers' awareness that really stiff price rises could add to U.S. inflation and to international financing problems, eroding the value of their dollars and foreign investments. But that is not to say that present conditions will remain as constant as the desert sun. Petromin, the national oil agency, could one day decide to bypass Aramco and market its oil in direct government-to-government deals on behalf of British, West German, French, and Japanese companies that agree to join Saudi Arabia in joint industrial ventures in the kingdom.

In January 1980, the Rand Corporation, a private research organization based in Santa Monica, California, that has close ties to the U.S. government, issued a disconcerting ninety-three-page study sponsored by the Defense Department. Its theme is that the Saudis' future oil income, large though it may be, still may not be large enough to cover their ambitious growth plans. The results could be serious inflation and balance-of-payments deficits as early as 1985. The third five-year development plan begun in May 1980 will cost nearly $300 billion. If this has to be modified, there will be many disappointed expectations, which could lead to the worst possible scenario.

Imagine the moderate, pro-American House of Saud replaced in Riyadh by a hostile mob of Moslem militants. Imagine the exclusive Equestrian Club, whose 185 or so members are all big business or royal family, overrun, its sun-shielding smoked glass smashed, its luxurious leather sofas overturned and trampled on, its $10 million clubhouse, its squash courts, and its swimming pool vandalized. Imagine poor Arabs handcuffing members of the House of Saud, brandishing automatic weapons at our diplomats and businessmen, demanding the ouster of all foreigners, and calling for a return to the austere way of life governed by the Koran and the traditions of the Prophet.

If this scenario ever comes to pass in Saudi Arabia, in the United States oil will sell for $70 a barrel, gasoline will be $5 a gallon under strict rationing, homes will be unheated, factories will close down, airline flights will cease, and there will be a run on international banks like the Chase Manhattan and on the stock markets to unload the shares of the four Aramco partners. It is, of course, only a possibility, but it seems far less fantastic since the Iranian revolution. It has finally sunk in on us that the United States and the rest of the industrialized world will continue to be vitally dependent on the Persian Gulf nations, and especially Saudi Arabia, through the end of the petroleum age.

The Bear Has an Appetite

What about the Russians, who produce all of their own oil? Are they that much better off?

Since 1974, the Soviet Union has been the world's largest oil producer. Of its 1980 average production of about 12 million barrels a day, it exported about 3 million barrels, half to its allies in Eastern Europe at "friendship prices," 35 percent to Western Europe, and the rest to friends like Cuba and Vietnam. Oil exports are the Soviet Union's principal means of earning foreign exchange.

In 1980, Saudi Arabia, the second leading producer but the biggest exporter, drew about 9.5 million barrels daily. Saudi Arabia has larger proven reserves than the Soviet Union. Although Soviet oil reserves are a state secret, experts think that explored reserves—which include proven, probable, and some possible reserves—come to about 75 billion barrels, compared to Saudi Arabia's 150 billion barrels of proven reserves.

Lately the Soviet Union's ability to continue as the world's leading producer and a major exporter has been called into serious question. During World War II, the Soviet republic of Azerbaidzhan, with its fields around Baku, produced most of the Soviet Union's oil. Now, after nearly a century of drilling, oil production there is stagnating. In 1978, the production of Azerbaidzhan, which borders on Iran, accounted for only 3.5 percent of total Soviet output. There have also been declines in the older fields in the region between the Volga River and the Ural Mountains in European Russia.

Since the beginning of production there in 1964, western Siberia has become Russia's largest producer and its main hope to close the gap left by the depletion of older fields. But there have been development problems in the harsh sub-Arctic environment, where roads and power lines are scarce and prospective fields are widely scattered in swampy forests. It would appear that Russia, too, has an energy crisis. President Leonid Brezhnev, at a meeting of the Communist Party Central Committee in December 1979, struck a familiar note when he termed energy conservation "the most important national goal."

In December 1980, Petrostudies, a Swedish research company, announced the possible discovery of a giant new oil field in western Siberia called Bazhenov, with alleged reserves of 4.5 trillion barrels—one hundred times the Soviet Union's current proven reserves. However, international experts discounted the claim and Petrostudies offered no evidence to back it up. Since 1965, only one new giant field—Fedorovsk—is known to have been found in the west Siberian basin.

Back in April 1977, the Central Intelligence Agency issued a study, based on satellite photos of Soviet oil facilities and other intelligence reports, that concluded that Soviet oil production would reach its highest level in the early 1980s and then decline. The CIA reiterated its position in a study issued in August 1979:

> The growth of Soviet oil production has slowed sharply since the mid-1970s. Output may peak in 1979 or 1980 and then decline sharply. It has already peaked in all the major producing regions except Western Siberia and growth is slowing even there. . . . Over the next few years, the USSR must either cut domestic oil use sharply or reduce oil exports to Eastern Europe and the West.

Not everyone agrees with the CIA, and the agency has been known to make wrong estimates in the past. But assuming that Russian output will peak fairly soon—and the available evidence does point in that direction—isn't that good from the American and West European standpoint? Perhaps, but while it would further damage the Soviet Union's already fragile planned economy, it could have some ominous side effects.

If Soviet oil production declines, Moscow would, for the first time, have to obtain significant amounts of oil abroad, and this could lead to competition with the United States and its allies for access to the huge oil reserves in the Persian Gulf region. The competition could take the form of providing arms and military support to anti-Western organizations or governments in the Persian Gulf states in exchange for oil deliveries. In its most extreme form, it could lead to direct military intervention to secure the oil fields in neighboring southwestern Iran. Not long ago, most Americans regarded this as a highly improbable scenario on the premise that such military action would set off a shooting war, even a Third World War, and the Russians would not dare risk that. Since the Soviet drive into Afghanistan, however, this scenario seems much less improbable. After all, one of Hitler's major goals in attacking Russia was to take possession of the oil fields of Baku. Now that those once prized fields are running down, connoisseurs of irony theorize that the Soviets, faced with a voracious appetite for oil, will one day mount a military drive through Afghanistan and Iran to satisfy that appetite in the Persian Gulf.

Whatever the merits of this theory, it is indisputable that the Soviet Union needs advanced oil exploration and drilling technology to maintain and increase its current output. This poses an interesting dilemma for the White House and the State Department.

While the United States is no longer the world's leading oil producer, it is still the leader in petroleum technology, and since the early 1970s, Moscow has purchased almost $1 billion worth of that technology. American geophysical equipment can see as deep as 15 miles below ground; Russian equipment has a limit of 3 miles. The Russians, who have given priority to their military and space programs, need that American equipment, just as they need modern drill bits, pumps, and (especially) sophisticated offshore drilling technology. Capitalist manufacturers like Otis Engineering Corporation, Baker Trading Corporation, and Camco Inc., all Texas firms, have been delighted to find willing buyers for their wares among the communists, who are hamstrung by exploration and drilling techniques that are at least twenty-five years out of date.

Eleven American companies attended an oil and gas equipment exposition in Baku in 1979. Dresser Industries of Dallas has developed a $144 million hard-metal drill-bit plant for the Soviets that uses Western technology. The Armco Steel Corporation of Middletown, Ohio, sold a $50 million semisubmersible oil-rig platform to the Soviets for assembly in Astrakhan, a port in the Caspian Sea noted for its caviar. The Soviets have expressed interest in buying an entire American factory to produce such platforms.

This is the first part of the dilemma: Should the U.S. government permit the sale of such equipment to its great-power rival? Without it, the Soviets' ability to keep their economy going would be weakened and so would be their political control over the East European allies they supply with oil. In the summer of 1979, the Carter administration, displeased by the Soviet invasion of Afghanistan, put oil equipment and technology on the list of strategic items that required political clearance before they could be exported. Then, in December, the White House approved twenty-two oil-technology licenses, and in March 1980, permitted the sale of some equipment but embargoed technology that could be used to modernize the Soviet energy industry. This, it was hoped, would give Washington some leverage over Russia's future actions in the Middle East.

There are obvious dangers in such an embargo—which brings us to the second part of the dilemma: Continued American frustration of the Soviets' desire to maintain self-sufficiency could force them into buying oil on the open market, which would further drive up petroleum prices for American consumers. Even worse, it could provoke military adventurism to satisfy the appetite of the Soviets and their allies for oil. So perhaps it would be in American interests to aid the Soviets in modernizing their oil production capabilities and developing nuclear and hydroelectric power and their vast natural gas reserves.

This dilemma underscores the fact that today the international oil business cannot be separated from the larger considerations of international power politics. As long as oil remains a major factor in powering national economies and military machines, it will be bought and sold, consumed, haggled over, and perhaps fought over.

The Transition Time

Will the day ever come when there just won't be any oil anywhere? Yes, but the question of when depends on a variety of factors such as the size of new discoveries, production cutbacks by OPEC members, the effectiveness of conservation measures, political upheavals, energy/gross national product ratios, the slowing of economic growth rates, and demand.

The respected Petroleum Industry Research Foundation of New York City made a detailed study of the question and reported in May 1978: "To estimate even approximately the decade when oil resources may actually approach exhaustion is beyond our ability and, we earnestly believe, that of most other forecasters, given the vast interplay of factors on both the supply and the demand side which will determine this." Nevertheless, the foundation was willing to hazard an educated guess, based on a consensus of informed sources, that, at 1976 levels of production, remaining resources were equivalent to nearly eighty years' supply and concluded that "if production could be held to present [1976] levels, the resource would presumably last until around the year 2057."

More urgent than the question of when the oil will finally run out is the question of when its availability will begin to decline seriously and how much time will then be left for the transition to other forms of energy. Here the experts are in general agreement. Although discoveries in the North Sea and Alaska are already making substantial contributions to supply, and the large new discoveries in Mexico and the potential in the High Arctic and China may ultimately stretch the life span of the world's reserves for a few years, nothing now in sight changes the basic forecast of production peaking and beginning to slide in the 1985–1990 period. As the federal government's *Project Interdependence* report of June 1977 put it:

> If we plan on the basis of the consensus view of the Nation's leading geologists, the Nation will be better prepared if the decline sets in by the mid-1990's as projected by [M. King] Hubbert. If, on the other hand, substantial discoveries are made in the regions which . . . have been underestimated in terms of their oil and gas potential, the world will be pleasantly surprised. . . .

Walter J. Levy, the noted oil consultant to industry and governments, addressed the question in an article that appeared in the *New York Times* of January 4, 1979:

> We must thus assume that the period available to us for transition from an oil-based economy to one founded substantially on new energy resources will probably not exceed 20 to 25

> years. The need to provide for the world's medium and longer-term energy future is clear, but not the means of getting from here to there.
>
> Coal will certainly make a substantially increased contribution, but mainly in countries with extensive resources. Moreover, shale oil, tar sands and very heavy oils will become a significant factor only if there are technological breakthroughs. The prospects for solar, wind and wave power and for energy from organic substances must be evaluated as relatively marginal. Atomic energy is the one feasible source that could make a major contribution, but its development is stymied by the problems of proliferation, waste disposal, and political and public opposition.
>
> By the mid-1980's or early 1990's, when the specter of imminent oil shortages begins to haunt the world, the pressures for large advances in the real price of oil are almost certain to become decisive. We have already experienced the economic consequences of the 1974 oil-price explosion, and since 1974 the progress of the world economy has become dangerously vulnerable to the effects of future sizable advances in the cost of oil.
>
> The importing countries find themselves between the Scylla and Charybdis of a slow rate of development and a policy of sustained growth that would entail the danger of an early oil shortage accompanied by high oil prices. Either case would involve the risk of a recession and political instability.
>
> Oil is a finite resource whose depletion is in sight. The consuming countries thus must set as their most important goals the further reduction in the amount of energy used per unit of production, establishment of optimum conditions for the expansion of traditional sources, and the development of large, dependable new energy supplies.

The temptation upon reading such gloomy assessments is to regard them as exaggerated. After all, in the past, something always turned up whenever the experts warned us we were running out of fuel. Surely, something must turn up again to get us out of this bind.

It might, at least for a while, but the cost should give us pause.

Chapter Nine

The Last Frontier Is Underwater

There is one last frontier in the world of oil and natural gas—the outer continental shelf. The United States has 1.8 million square miles of petroleum-prospective acreage off its shorelines, of which less than 5 percent has been leased by the federal government to private oil companies. The operative word here is *prospective*, for no one really knows how much oil and gas can be produced by offshore drilling, although its advocates estimate that at least one-third of all future domestic production will occur offshore.

It must be remembered, however, despite the reassuring jargon about the whereabouts of petroleum declaimed by their practitioners, geophysics and geology are highly speculative fields. If they were exact sciences, all the oil and gas under the seas of the world would have been located already. Nonetheless, there are indications of offshore U.S. areas where sizable deposits might yet be found, as they have already been found in the Gulf of Mexico. The outer continental shelf accounted for 14 percent of America's 1979 oil production and 23 percent of its gas.

The search involves arcane technology and strange, very costly equipment. Continental Oil Company in January 1980 unveiled a new type of drilling structure, called a tension-leg platform, for use in the North Sea. The new platform should make possible the production of oil lying under 2,000 feet of water. The entire project, which Conoco is sharing with seven partners including the British National Oil Corporation and Gulf Oil, will cost about $1.4 billion, including the necessary pipelines.

The search also involves staggering sums of money for buying leases and exploration. That lets out the little guy and means that the job has to be done by the big oil companies, alone or in combination, with their massive cash flows. The oil companies obviously wouldn't be in this risky game unless they thought they could eventually make it pay, and have been critical of the Interior Department's slowness in auctioning leases. By contrast, Interior's critics contend that the department has been leasing the companies too many tracts without adequate knowledge of what lies under the sea bed and its real, long-term value. Such critics include the General Accounting Office; Senator Henry Jackson, the Washington Democrat who headed the Senate Energy and Natural Resources Committee in 1980; and Dr. H. William Menard, director of the U.S. Geological Survey. Dr. Menard said in 1977: "It would be illegal to act with such scanty information in leasing any other mineral than oil and gas."

How are the big oil companies going about the exploration of America's last frontier, and what success are they likely to have? To answer that question, which has a critical bearing on America's energy situation, let's first examine the experience of Texaco and its foray into the Baltimore Canyon in the Atlantic Ocean, 100 miles east of the boardwalk and gambling casinos of Atlantic City, New Jersey. This was a high-risk adventure, not just because of the sums of money wagered on its potential success, but because it could create environmental havoc far beyond what is possible with conventional onshore drilling.

To understand why Texaco was willing to take a considerable risk in the Baltimore Canyon, it is necessary to journey backward in time about a third of a billion years to the Mississippian Age, when the world's continents were drawing together to coalesce, eighty million years later during the Permian Age, into one supercontinent, Pangaea. The sea level was then very low and massive continental glaciers covered Pangaea. Over millions of more years, the supercontinent broke apart, the glaciers melted, and the sea rose. Finally, the Atlantic Ocean was born as the gulf between the African and American land masses widened. Ten million years ago, by the Miocene Age, the shorelines of the Atlantic were lapping against the inner edges of America's present continental shelves.

Gradually the depth of the oceans sank again with the buildup of huge glaciers. About twelve thousand years ago—during the Ice Age—the surface of the Baltimore Canyon lay exposed above the land mass, covered by rivers, forests, meadows, and ponds. Men using fluted spears hunted the woolly mammoth, musk ox, and giant moose that roamed over the rich soil with its lush vegetation. Then the earth turned warmer, the glaciers melted again, and the Atlantic Ocean rolled over most of the Baltimore Canyon.

This structural depression or trough we call the Baltimore Canyon is filled with the sediments, crushed fauna, flora, and animal bones that typically are transformed into the petroleum hydrocarbons that can eventually light up your kitchen stove or put a tiger in your tank. However, there is only one way to prove that commercial quantities of oil and gas are trapped in this prehistoric underwater canyon, and that is by drilling.

In 1962, Texaco began surveying the structural depression along the shores of New York, New Jersey, and Connecticut with seismic survey ships. Sonic waves penetrated to the various sedimentary rocks that lie beneath the floor of this part of the Atlantic Ocean, then vibrated back to the ship to give a map of the structure of the underwater earth formations. Rocks were brought up and compared to rocks from oil-bearing sites in the North Sea. In all, Texaco accumulated 28,000 miles of seismic data.

In 1966, Texaco joined with other oil companies in the annual mapping of geologic profiles and shared the data; by 1971, potential hydrocarbon-bearing areas in the Baltimore Canyon were as well defined as they would ever be. A sealed-bid lease sale, in which companies alone or in combination compete for the right to explore for oil and gas on specific tracts, had originally been scheduled by the federal government for 1969, but it was postponed because of a dispute between the states and the federal government over ownership of offshore mineral rights. Litigation by environmental groups, which objected to offshore drilling, further held up exploration. Though the environmentalists were usually dismissed as eccentric idealists by the oil companies, they were taken more seriously in most quarters because of the ravages caused by the Santa Barbara blowout in 1969 (still to come was the disaster caused on Texas shorelines by the blowout in Campeche Bay, Mexico, in 1979). Eventually, the environmentalist suits were resolved (although the potential chaos posed to East Coast beaches by an Atlantic Ocean blowout was not), and in 1975 the Supreme Court upheld the federal government's right to lease this part of the outer continental shelf. A sale date of August 17, 1976, was announced.

Now Texaco and other major oil companies that had been making geochemical analyses of the area were faced with an economic decision. How much money should be risked against the *estimated* value of the *potential* reserves? And what specific tracts should the money be bid on? Whose geologists were right in this guessing game? Wrong guesses could result in the loss of millions of dollars.

Texaco concluded that the investment and risk were too great for the company to make the Baltimore Canyon "play" alone, so it became the operator for a group of six companies with varying interests—a good illustration of the complex financing required by such huge projects. The Texaco group submitted to the Bureau of Land Management a sealed bid of $309,541,000 for nineteen tracts. It won two of those tracts at a cost of $50,610,000. All told, oil companies—including Exxon, Mobil, and Gulf as well as Texaco—invested about $1.1 billion for leases off the East Coast, perhaps the biggest commitment of dollars for an exploratory venture by private enterprise in American history.

The justification for such a massive investment lay in the enormous potential for profit. According to estimates by the federal government, the oil and gas reserves beneath the waters of the outer continental shelf are worth from $200 billion to $1 trillion at 1980 market prices. The American Petroleum Institute has estimated that from 1953 to 1977 the petroleum industry paid out about $25.4 billion in offshore bonus bids, royalties, and other payments to the federal government. The value of the production derived from this investment was $33.3 billion (there are oil men who contend that this is not an undue rate of return).

Despite its lengthy prospect analyses, stratigraphic planning, and global tectonic analysis, Texaco would not know what was in the two tracts it had won until it began "spudding in"—or drilling—through the layers of sedimentary rocks in the trough. Offshore drilling was an old story in the Gulf of Mexico, of course, where some twenty thousand wells had already been drilled, but no productive wells had ever been drilled in the Atlantic. The instrument chosen to discover if there were commercial quantities of oil and natural gas under 432 feet of water in Block 598 was one of the strangest-looking machines ever devised by mankind.

The *Ocean Victory* was a semisubmersible vessel self-propelled by two electric submarine motors powered by diesel engines on its main deck. It had four 28-foot-diameter parallel pontoons and twelve 100-foot-high legs and was strong enough to withstand hurricane winds. The *Ocean Victory* was as long as a football field and nearly as tall as a thirty-story building when its drilling rig was raised. Three pilots, or captains, shared the watch and navigated the craft from a tiny wheelhouse.

As the *Ocean Victory* slowly made its way north 12 miles off the coast of Miami Beach, some pleasure boats radioed the Coast Guard that they had spotted an object from outer space. But the drilling rig had only come from the Gulf Coast of Louisiana, where it had been built by ODECO (Ocean Drilling and Exploration Company) and leased to Texaco. The cost to Texaco, including helicopter service from Atlantic City and supply boats from a base in Davis, Rhode Island, was about $60,000 a day.

After an uneventful journey of 1,500 miles from the Gulf to the Baltimore Canyon, the *Ocean Victory* dropped 30,000-pound anchors marked by pennants from each of its four outermost legs and its crew set to work. They numbered sixty-five. There were toolpushers, responsible for drilling operations, welders, roustabouts, anchor crews, and roughnecks—all supervised by Texaco's own managers. The crews worked two-week tours of duty, then were relieved for two weeks by other crews flown in by helicopter from Atlantic City. On-tour amusement was restricted to sleeping, eating, playing cards, and watching current movies on video cassettes. Most of the crew were "good ol' boys" from Texas and Louisiana who were concentrating on "reachin' and gittin' it" and "makin' hole."

On the late afternoon of Sunday, April 16, 1978, the drilling derrick was raised, a drill string was lowered into the water through the rotary, which revolved and powered the 17½-inch-diameter blue-painted drilling bit attached to the bottom of the string. At a command from the driller's console, the rotary turned at 5:10 P.M. and the rock bit took its first bite into the unknown, monitored by underwater cameras.

The drill turned round the clock for days, weeks, months, as the crew worked, slept, watched movies, and hoped. Then, on the balmy day of August 14, 1978, at a depth of almost 16,000 feet, a roar of flaming gas came out of the wildcat well—the first discovery of hydrocarbons off the U.S. East Coast. Though this was encouraging, it did not mean the search was over. Other wells would have to be drilled nearby to confirm that this was a commercial reservoir worth developing.

Drilling was begun on a second well 1½ miles away. Autumn faded and the wintertime Atlantic weather closed in. The *Ocean Victory* heaved, dipped, and yawed as it was pounded by 25-foot waves; winds of up to 50 miles an hour howled across its derrick; the freezing cold and snow caused fifteen crewmen to quit and head south.

The drill bit had reached 17,708 feet by early March 1979 with no results whatever. On March 12, Texaco announced that it had fully tested the well "without encountering oil or gas in producible quantities." It was a dry hole, as are 90 percent of exploratory wells; the company plugged and abandoned it.

The high hopes for the Baltimore Canyon were now fading. Already on December 27, 1978, Mobil had announced that it was abandoning its first well after drilling into the top of the

Baltimore Dome. This is a large geological formation of rock bulging upward from the sea floor. Such domes have the potential of trapping oil or gas as an upside down bowl can trap air in a bathtub full of water, but Mobil found "no significant quantities" of hydrocarbons in this one.

After drilling three dry holes, Shell announced on February 14, 1979, that it was giving up; it had spent about $90 million on leasing and drilling. On April 2, Gulf abandoned its second exploratory well 80 miles offshore of Atlantic City and said that it had "no immediate plans for additional drilling." Exxon drilled four dry holes and suspended operations. That left Texaco as the only company drilling off the mid-Atlantic in 1979.

Texaco began drilling its third well 1.1 miles northwest of a discovery of oil and gas reported by Tenneco Inc. in May and June, and just within the boundary of a 9-square-mile lease tract owned by a group headed by Tenneco and including Sun Oil and R. J. Reynolds Industries. The latest well would cost an estimated $19 million, and this cost was being shared by the Tenneco group and the Texaco group. Texaco held a 31.5 percent share in the latter, the remaining 68.5 percent being held by Getty Oil, Sun Oil, the Allied Chemical Corporation, the Transco Companies, and the Freeport Minerals Company.

On October 22, the *Ocean Victory* struck natural gas at a depth of 15,699 feet. On November 12, the rig encountered more natural gas in the same hole, and on November 23, it made another find at about 13,000 feet, following the usual custom of making flow tests from the bottom up.

It was certainly good news, but Texaco cautiously did not say that its finds made production feasible, only that it was plugging the joint Tenneco well and moving its leased drilling rig to test yet another confirmation well on its original tract. No one appeared to commit the big sums of money needed to finance the production platforms, pipelines, and other equipment essential to bring natural gas ashore.

As 1979 ended, it became evident that while Texaco had fared better than most, the Baltimore Canyon had only underscored the capriciousness of hunting for petroleum. Instead of the expected billions of barrels of oil, the U.S. Geological Survey revised its estimates downward to about 4.1 trillion cubic feet of recoverable natural gas and 600 million barrels of recoverable oil off the mid-Atlantic. These figures would hardly give the Saudis pause. They meant that the area could supply all of America's oil needs for only about a month and its natural gas needs for about two months. "If there were a lot of oil off Jersey," Dr. Kenneth Emery, a geologist at the Woods Hole Oceanographic Institute, commented in 1980, "it would have been found already."

Even when offshore drilling is successful, the usual lag between discovery and production is seven to ten years, so even a stupendous find would not produce a near-term cure for America's energy problem. Yet the offshore search has to be continued, for that is where the future action lies.

The Gulf of Mexico, with its mild weather, familiar geology, and relatively soft geological layers, is the star of the expensive offshore oil business. It is known to oil men as a "mature" province, meaning that the largest discoveries have already been made there and whatever remains will be more difficult to find and will require more sophisticated technology. Yet the Shell Oil Company (and fourteen partners) have enough confidence that the Gulf is not over the hill to have invested $800 million in the Cognac Project (a code name) 15 miles from the mouth of the Mississippi River. This field is estimated to contain 500 billion cubic feet of natural gas and 100 million barrels of crude oil. To recover them, Shell has spent $265 million on the Cognac Platform, which was completed in the fall of 1978. It is a twin-derricked gray-and-yellow giant with most of its 1,025-foot legs sunk under the deep-green water off Louisiana. Cognac is the world's biggest drilling platform—taller than the Empire State Building's observation deck—and the first one able to operate in water deeper than 1,000 feet. If Shell's strategic planners are right, the Cognac field will produce about 150 million cubic feet of gas and 50,000 barrels of oil a day by the beginning of 1983.

Even the Cognac field is not going to solve America's oil problem, however. By the 1990s, most, perhaps all, of its production will be natural gas. This simple fact underscores a major change in the petroleum business in the older provinces. The first big finds there, both in the Gulf of Mexico and on the mainland, were oil. Because of lighter geopressures, it could be found by drilling relatively shallow wells. Even when natural gas was discovered at the same time, its selling price was so low that producers often declined to exploit it. As the shallow wells became depleted, however, producers had to drill deeper to get any oil, and at these levels, gas was more likely to be found. As the price of natural gas rose, it became profitable to produce and sell it. This illustrates yet again that where a dollar can be made, the industry will manage to harness its expertise and funds to go after it.

Wherever it can be used as a fuel by homeowners and adapted for use by heavy industry, natural gas is the most attractive alternative to imported oil. Still, it can never replace crude oil as a source of gasoline, aviation jet fuel, marine fuel, and lubricating oil. Crude oil production in the Gulf of Mexico, which began in the late 1940s, has been declining since 1971, and new fields like the Cognac will not reverse that decline. They only constitute a holding action until the industry can find substantial offshore "virgin" provinces elsewhere.

Always in search of profits, the industry has tried hard, with varying results. A group of companies headed by Exxon paid the federal government $632 million for the right to drill into the Destin Dome off the west coast of Florida and found almost nothing. Substantial finds were made off Santa Barbara, the picturesque California resort community of Spanish-style buildings and gleaming beaches, but this good luck turned into a nightmare.

On January 28, 1969, a runaway gusher erupted on Platform A, 5 miles offshore. Before it was finally plugged, the blowout spewed more than 10 million gallons of crude oil that fouled the ocean and beaches for 20 miles up and down the southern California coast and put all offshore drilling under a cloud as black as the oil

It later became obvious that the blowout reflected undue

government and industry haste to exploit the Santa Barbara channel. A lid was clamped on development there, which was only gradually relaxed after the Arab embargo of 1973–1974 with the provision that drilling conform to the new National Environmental Policy Act. This act requires advance study of the impact of any project in which the federal government is involved. But lease sales moved slowly because of state and local resistance and litigation.

By 1980, there were only fourteen platforms in the channel. Seven of them were on federal leases and the other seven were within California's 3-mile zone. Drilling had produced less than 15 million barrels the previous year. That was less than one day's national consumption. Santa Barbara had regained its affluent serenity through the efforts of a civic group called GOO (Get Oil Out), which continued to campaign against all offshore leasing. Nevertheless, the federal government planned two lease sales of offshore tracts scattered from the Mexican border to northern California.

Undoubtedly the most promising offshore area still to be explored is in the Beaufort Sea off the shores of Alaska. Geological surveys suggest that this is the last of the great undiscovered oil fields in American territory. Estimates of how much oil could be found off Alaska's outer continental shelf range up to 25 billion barrels—nearly three times the proven onshore reserves of Alaska's North Slope. Federally owned lands in the Beaufort Sea, 275 miles above the Arctic Circle, are only the most promising of many in this frigid region. The Canadians have drilled in their sector of the Beaufort Sea since 1977. One company, Dome Petroleum, brought in a 20,000-barrel-a-day well in the fall of 1979. It was the biggest strike ever made in Canada.

In the American sector of the Beaufort Sea, the stakes are enormous—for the oil companies, for the American people, and for the Eskimos who live there in an unspoiled environment. In December 1979, oil companies bid over $1 billion for the privilege of drilling in 514,191 acres of the Beaufort Sea, undaunted by the storms, temperatures as low as -60° F, and a slow-moving ice sheet up to 15 feet thick that covers the sea in the dark winter months. If these shallow waters yield the estimated 1 billion barrels of oil and 2 trillion cubic feet of natural gas—all with easy access to the already existing pipeline—that would come to more than $40 billion worth of energy at 1980 prices.

As usual, there are formidable obstacles. The natives, as well as environmentalists, are convinced that the oil companies' technology of drilling and cleanup in these icy waters lags far behind their appetite for profits. They fear that a blowout would damage rich fisheries, deposit a permanent residue of muck along a virtually untouched wilderness coast, and endanger the bowhead whale, whose meat is a staple of the Eskimo diet. The coastal hamlet of Kaktovik, populated by 175 Eskimos, and several other Alaskan groups have brought suit to halt exploration. Their lawyers contend that if the Eskimos believe that drilling will disturb the whales, which migrate through the Beaufort Sea, they could be afflicted with "serious mental and emotional anxiety." The National Wildlife Foundation and other groups have also brought suit to ban drilling to protect the whales and the environment.

While awaiting resolution of these suits, the Interior Department showed enough concern to restrict drilling to from five to seven months a year, starting in November, when the bowheads are locked out of the Beaufort Sea by ice. The government has also required the oil companies to test their rigs for two years before operating them in water more than 42 feet deep in order to determine if the rigs can withstand the moving ice packs, which could knock them over and cause spills.

The oil men, of course, feel deeply frustrated by these delays and routinely ridicule environmental suits and government regulations. They would like to be given full access to all offshore Alaskan areas by the end of 1982, but they will be lucky if the Interior Department completes lease sales by 1985. Besides the Beaufort Sea, five other huge undersea basins are involved. U.S. Geological Survey scientists believe, in the words of Dr. Charles Masters, that they "may have enormous petroleum resources."

Three of these basins—the Hope, St. George, and Norton—lie

north of the Aleutian Islands in American waters. The other two—the Chukchi Basin in the Chukchi Sea and the Navarin Basin in the Bering Sea—lie off the sparsely populated west coast of Alaska. These, the most promising of the group, unfortunately extend into waters that might eventually be claimed by the Soviet Union because of the vagueness of the boundary lines set by the 1867 treaty by which the United States bought Alaska from Czarist Russia.

The U.S. Geological Survey discovered all of these basins and informed the oil companies about them. "Now they're up here thick as fleas doing exploration," Dr. Masters said, but cautioned that commercial production is many years away. "Production problems will be severe," he pointed out, "since drilling in these areas would stretch our technology to the limit." The oil companies are confident that they have now, or can develop, the necessary technology. Some of it is as exotic as any used in the aerospace industry and as unfamiliar to the public as a bowhead whale.

For all their carping about environmentalists, Eskimos, and dawdling government bureaucrats, the oil men's biggest enemy is ice. Shifting, grinding Arctic pack ice is capable of crushing an oil rig in seconds, causing spills that would be especially difficult to locate and clean up if released under the ice in winter. To counter this danger to conventional offshore rigs and platforms, the oil companies have developed man-made islands of rocks and gravel taken from the North Slope, which they insist are strong enough to ward off the treacherous ice. Rigs are mounted atop these artificial islands and drill through them to the ocean floor. Environmentalists want to know what will happen when the ice buckles, bunches, and rears up to moving 100-foot ridges within seconds.

Then there are the icebergs. The Canadians working in "Iceberg Alley" off the northern coast of Labrador have developed a technique of stringing cable around threatening icebergs. Two ships then pull the line taut and maneuver the berg away from the drill site.

Oil men in the northern regions are on constant alert for ice.

They have developed a type of radar that can detect ice from the sky through cloud cover and darkness, and another that gives a profile of the bottom of an ice pack. These are fascinating innovations, but none is guaranteed to prevent oil spills. The effects of an oil spill in these regions simply cannot be predicted. Neither can the disruptions that land-based facilities might cause Alaska's fragile ecological balance, nor how much oil will be discovered offshore of Alaska. If some 25 billion barrels of oil are found, it would be enough to satisfy one-fifth of America's needs for fifteen years. Under current production difficulties and federal leasing schedules, however, little of that oil will be available to consumers before 1995.

On the opposite side of the North American continent, oil companies are working less hostile waters. In January 1980, Chevron Standard Ltd., a subsidiary of Standard Oil Company of California (Socal), announced that its Hibernia 0-15 well could one day produce 20,000 barrels a day. Hibernia lies 205 miles east-southeast of St. John's, Newfoundland. Here there are few icebergs, although one of them did sink the *Titanic*. There is no pack ice, so drilling can go on the year round. The weather is hardly inviting, but the cold, wind, and rain off Newfoundland are comparable to conditions existing in the North Sea, where pioneering techniques with platforms suitable for such weather were first perfected.

Hibernia 0-15 is only the first well in a 525,000-acre block that could yield other big strikes. Like other such undertakings, this one is too big a financial gamble for any one major oil company, let alone even the biggest of independents. Socal is only the operator for a group that includes Mobil (28 percent share); Gulf Canada Ltd. (25 percent); Socal (18 percent); Petro Canada, which is the Canadian national oil company (25 percent); and Columbia Gas (4 percent). The payoff for this group, and supplies for consumers, will be a while coming. Informed estimates put commercial production at Hibernia 0-15 from five to ten years off.

Some 800 miles to the southwest, Mobil Canada reported in 1979 a natural gas strike near Sable Island, 130 miles off Halifax. Still farther south, in American waters in a West Virginia–sized patch of the Atlantic continental shelf, is the Georges Bank off Cape Cod, where the U.S. Geological Survey estimates that 123 million barrels of oil and 870 billion cubic feet of natural gas are trapped in geological structures beneath the sediment. Oil men are avid to get at that oil and gas. The only counter to their desire is the fact that the Georges Bank already has a proven resource that, unlike oil, is renewable. It harbors a cornucopia of some two hundred species of fish and supports a $1-billion-a-year fishing industry that provides 15 percent of the world's annual catch.

An alliance of environmentalists, fishermen, and the Massachusetts attorney general fought a bitter legal battle all the way up to the Supreme Court to stop the Interior Department from selling tracts to the oil companies. The alliance feared that a major blowout—and the continuous discharge from day-to-day operations—could devastate the rich fish-spawning area, which it wants declared a marine sanctuary. Twice an auction was postponed as a result of litigation brought by the Conservation Law Foundation and the Massachusetts attorney general. Oil men contended that a blowout was not inevitable, that parts of the Georges Bank were only 9 feet deep and that the greatest depth was but 300 feet, that safeguards were adequate, and that the Commonwealth of Massachusetts was being hypocritical in accepting the benefits of offshore drilling in other domestic areas while declining to take any risks with its own. As usual, given the national demand for energy and dependence on vulnerable imports, the oil companies won the battle.

The auction of 660,000 acres (with spawning grounds excluded) was held on December 18, 1979, in the Rhode Island Veterans Auditorium in Providence. It was attended by Interior Department officials, two hundred oil company executives, and a large contingent of state troopers. When the sealed bids were opened, the federal government had high bids totaling more than $800 million for drilling rights in 116 tracts of 9 square miles each.

The big winner was Mobil, which submitted high bids in combination with several partners for sixteen tracts valued at $494.3 million. The highest bid on a single tract was $80.3

million, made by a consortium of Mobil, Amerada Hess, and the Transco Exploration Company. The auction was enlivened by three protesters who got into the balcony and dropped leaflets on the heads of the oil executives below. They managed to toss a dead fish onto the stage and hurl at least four oil-filled plastic bags at the executives before being ejected by state troopers.

The reaction of the fishing community, some of whose families have worked the Georges Bank for 350 years, was summed up in a comment made by Angela Sanfilippo, president of the Gloucester Fishermen's Wives Association: "It's a disgrace. There are no safeguards that we can see. This is a threat to our livelihood. It is a failure of our Government. After so many people appealed to them, they went ahead anyway."

It is impossible to estimate how much oil will be found in the Georges Bank until the oil companies begin spudding in. It is equally impossible to know whether exploration will result in serious damage to the beaches of Cape Cod, Nantucket, Martha's Vineyard, and Long Island, and to the haddock, cod and yellowtail, scallops and lobsters, squid and swordfish that thrive in the converging Gulf Stream and Labrador currents in one of the richest fishing grounds in the world. The oil companies maintain that any danger is minuscule. They point out that of the 292 million barrels of oil produced from offshore U.S. wells in 1978, only two spills exceeded 50 barrels—the largest being 135 barrels.

Yet the danger is real. While offshore drilling has the potential to produce enormous quantities of oil, it also has the potential to create a disaster for the environment and for the men who follow the little known trade of "makin' hole" from offshore drilling rigs. Most Americans find it difficult to summon up concern for the Gloucester fishermen, the Eskimos, the bowhead whale, and a pristine Alaskan wilderness they'll never see. But a brief reminder of a few of the incidents that have taken place in the Gulf of Mexico might provide some reason to ponder the trade-offs involved in exploiting the last oil frontier.

On March 24, 1980, a crew of forty-one was drilling for natural gas in 310 feet of water off Galveston, Texas. Their platform was owned by twelve oil companies and operated by Pennzoil. Just before dawn, the rig began to "kick," mud sprayed from the well, and the rig exploded before any of the crew could escape to safety. Before the fire burned itself out that same night, two men were dead, twenty-nine were injured, and four were missing in the high winds and waves that hampered rescue operations. "The concussion just blew me over the handrail," said one survivor, Stan Riley. "I was lucky I got blown off. The other guys on the production platform got burned." Many of those injured had jumped overboard from as high as 125 feet above the water, which is comparable to landing on a concrete highway.

On December 8, 1977, a helicopter crowded with workers slammed into a rig off Lafayette, Louisiana, and sank beneath 10-foot waves. Seventeen men died.

On April 16, 1976, a rig called *The Ocean Express* was being towed 30 miles off Port Aransas, Texas, when it foundered in a storm. Seventeen men died.

A "freak sea" with mountainous waves capsized a rig off Louisiana in June 1965, drowning one man. Eleven others survived, five of them by huddling in air pockets in the overturned rig for more than twenty-two hours.

The worst offshore disaster of all time did not take place in the Gulf of Mexico, but in the North Sea. The *Alexander Kielland,* anchored halfway between Scotland and Norway, was leased from Stavanger Drilling, a Norwegian company, and operated by Phillips Petroleum, the major contractor for oil and natural gas exploration in the rich Edda field. It was not a drilling rig, but a pentagonal 10,105-ton floating dormitory built in 1976 by the Paris-based Compagnie Française d'Enterprises Métalliques. Outfitted with sleeping quarters, mess hall, and movie theater, it housed 212 oil workers, predominantly Norwegians. These rotating crews lived there for two weeks at a time while operating nearby North Sea drilling rigs 240 miles west of Stavanger, Norway. The semisubmersible "floatel" was mounted on five columns that rode on pontoons in the water and were attached by cables to the seabed. It was touted as invulnerable to the weather and had often ridden out severe storms. But at 6:30

P.M. on the night of March 27, 1980, 65-mile-an-hour winds developed and 25-foot waves battered the platform. One of its legs snapped and it listed 45 degrees within seconds. Within twenty minutes the floatel capsized, suspended 100 feet beneath the North Sea from its four remaining pontoons.

Most of the workers were in the mess hall enjoying smoked salmon or steaks when there was a loud crack, the lights went out, and the alarm sounded. About forty were watching a western in the movie theater. The fortunate managed to scramble into lifeboats and dinghies. Others were thrown into the near-freezing wind-whipped North Sea. Most of those in the movie theater were trapped as the floatel sank beneath the waves.

Norwegian, British, Danish, Dutch, and German planes and helicopters and a flotilla of at least forty-five ships rushed to the scene, where they fought poor visibility, heavy seas, and winds to pull some survivors from the sea and the boats. Divers hoped that more survivors might be trapped in life-saving air pockets in the theater and living quarters, but tapping on them brought no response. Forty-eight hours later, the Norwegian government called off the search and said that a total of 123 men had died. No Americans were among them, although Phillips Petroleum executives had often visited the platform.

Investigators concluded that a crack beginning in a 14-inch hole drilled in a leg brace to hold a hydrophone led to the disaster. Hydrophones pick up signals from the sea floor and enable engineers to position the rig overhead. The crack had spread from the hole, causing one of the 1,000-ton legs to give way and the entire structure to roll over. But the investigators could not determine why the crack had occurred in the 14-inch hole, or how to prevent a recurrence. "What we don't know," said John Mihms, the engineering manager in Norway for Phillips Petroleum, "is why this hole behaved differently from others. Was there a faulty weld? A notch in the hole? A dent in the metal adjacent to the hole? We're still trying to figure that out."

China is a relative newcomer to offshore drilling, but the Chinese are now familiar with the problems it can cause. In

August 1980, the Chinese government indicted four oil industry supervisors on charges of criminal negligence in the loss of the Bo Hai No. 2 drilling rig on November 25, 1979. The Japanese-built rig had collapsed and sunk in stormy seas in the Bo Hai Gulf southeast of Peking, costing seventy-two lives and a $25 million investment. The unusual indictments were part of a campaign to demonstrate that no one is above the law in China. "Isn't it a common conception that top officials are like 'tigers whose backsides no one dares to touch' and that 'bureaucrats shield one another'?" Li Weihan, adviser to the Standing Committee of the Chinese People's Political Consultative Conference, asked. "This decision shows that whoever violates the law and discipline will be punished." The State Council, or cabinet, issued a "demerit first grade" to the deputy prime minister in charge of oil development for mishandling the disaster and said that the Petroleum Ministry had violated safety rules and tried to evade responsibility for what had happened.

All the same, offshore drilling has continued in Chinese waters, just as it has in American waters, along with attendant disasters. On August 30, 1980, the *Ocean King* drilling rig exploded and burned in the Gulf of Mexico 30 miles off the Texas coast, killing two men and injuring six.

Offshore drilling is not only dangerous for people, but also poses grave dangers to the environment. One warning about environmental dangers was served in an area of the Gulf owned by Mexico.

Since 1938, when Mexico kicked out American oil companies, the Mexicans, with substantial help from American technicians and equipment, have been running their own show. It was not much of a show until 1972. Then it turned spectacular when geologists, drilling among the rundown cattle farms, skinny children, and cactus-studded wasteland of Tabasco State, found the huge Reforma oil and gas field. Since 1972, Mexico has struck one enormous petroleum deposit after another, particularly in the Gulf, where its section is geologically similar to the waters off Texas and Louisiana that have been so productive.

Most of the drilling at more than fifty offshore wells has taken place in the Gulf of Campeche, Mexico's richest shrimp-producing region, where repeated reports of oil seeping from the sea floor by shrimp fishermen led Pemex, the state oil monopoly, to drill. There was little of the hopeful guesswork and frustration encountered by American oil men in the Baltimore Canyon. Pemex was sure that just one of the exploratory wells, Ixtoc-I, located in shallow water 58 miles northwest of Ciudad del Carmen off the Yucatán Peninsula, contained reserves of at least 800 million barrels.

On June 3, 1979, Pemex was drilling to exploit this bonanza when something frightening happened at Ixtoc-I. No one is quite sure, or willing to admit publicly, how or why it happened, but there was an unexpected loss of drilling mud in the hole and a torrent of oil, with its accompanying natural gas igniting on the way out, began spewing into the Gulf. The Mexican rig crew was unprepared for the blowout but reacted by using the procedures for capping a blowout they had learned from American experts. The main safeguard against any blowout is a mixture of chemicals, clay, and water known as drilling mud. The mixture is continuously circulated from the bottom of the well to the surface, where it is supposed to hold down any oil or gas that is hit. If the drilling mud fails, blowout preventers—a series of valves—are supposed to slam shut over the top of the well.

At Ixtoc-I, both the first and second lines of defense failed—or were not worked correctly. Under pressures never before met in the Gulf, oil shot out and began polluting the water at a rate of 30,000 barrels (1.26 million gallons) a day. Ten months later, Ixtoc-I was still running wild, although the rate had been reduced to 10,000 barrels a day. The Mexican government, with the help of a small army of foreign experts, made a strenuous effort to cap it, but the spill became the biggest in history—bigger than that caused when the Liberian supertanker *Amoco Cadiz* cracked open in the English Channel in March 1978 and lost more than 1.6 million barrels.

The results were of a weird nature never before seen on planet Earth—on land or sea. Oil slicks, tar balls, the brown mixture of mud and oil known as "mousse," and oil mixed with seawater into a kind of petroleum fudge appeared and drifted toward the Texas and Florida coasts. By early August, a sheen of oil that made the sea shine floated 600 miles north to the Texas Gulf coast to coat 150 miles of beaches between Brownsville and Corpus Christi with waves the color of chocolate milk.

Despite weeks of warning, the American team of experts and equipment from the Navy, Coast Guard, Environmental Protection Agency, Fish and Wildlife Service, and the National Oceanic and Atmospheric Administration had little success in stopping the oil from crashing onto the coast. That's because oil is much easier to contain at the well than it is at sea or on beaches. "Not a lot can be done at sea," said John Robinson, science coordinator of the American team. "You've got to stop it at the well site if you're going to stop it."

To stop it at the well, the Mexicans hired the redoubtable Red Adair and his team of blowout experts, and engineers from Brown Root in Houston. Not even they could cap this runaway. The Mexicans were, predictably, accused of bungling and malingering by some, but they did hire the best foreign experts and were willing to try everything. They drilled two relief wells and dispatched skimmer ships to sop oil from the water's surface and planes to spray chemicals that disperse oil. But Ixtoc-I, like some angry ancient Mayan god, would not be quenched and its continued eruption brought a political fallout.

The Carter administration, alarmed by a hefty cleanup bill (an average of $75,000 a day, which could eventually balloon to $4 million), diplomatically requested the Mexican government to contribute $3 million toward the cleanup. President José López Portillo angrily rejected the request, saying that it had no basis in international law. Aware that Mexico has proven reserves of 45 billion barrels of oil easily accessible for the American market and is not an OPEC member, the administration gingerly backed off.

The political fallout took a seriocomic turn in Texas, whose Republican governor, W. P. Clements, dismissed the spill as "much to-do about nothing" and advised the Mexicans that it would be foolhardy for the United States or Texas to sue to recover damages. The attorney general, Mark White, a Democrat, responded angrily that it was up to him and not the governor to decide who would or would not sue. Then it developed that Governor Clements was the founder of the Dallas-based company, Sedco, that owned the rig that had drilled Ixtoc-I. In his defense, Governor Clements protested that Sedco had turned the rig over to a Mexican company under contract to Pemex, and besides, he had dissociated himself from the firm upon becoming governor and had placed his interest in it in a blind trust. The fact that his son was Sedco's president had no bearing on his opinions, the governor claimed.

The Texas Gulf coast beaches escaped serious pollution only because of the weather. Changing winds and currents blew the mousse, fudge, and tarballs out into the middle of the Gulf of Mexico, where they floated around through the winter of 1980. Ixtoc-I gushed on uncapped, with an impact on marine organisms that has not yet been determined, although the flow was reduced from 30,000 to 10,000 barrels a day in August 1979 when thousands of canvas plugs and tennis-ball-sized lead and steel balls were pumped at high pressure into the pipe. Finally, on March 24, 1980, Petroleos Mexicanos capped the well by inserting cement plugs.

According to Mexican figures, 3.1 million barrels of crude oil were released into the Gulf during the ten months of the blowout, although some American experts estimate the total as four times higher. The exact whereabouts of the oil that has not washed up on beaches or been collected from the surface of the water remains a mystery (oil can float below the surface or sink to the bottom as well as float on the surface).

The cost of the blowout, including the value of the oil and the drilling rig, has been estimated at $225 million. Lawsuits totaling $337 million have been filed against Pemex and Sedco by south Texas resort and fishing communities, but so far the Mexican government has refused to acknowledge the lawsuits, which has not improved relations between Mexico and the United States.

It is hard to put a dollar figure on the cost to the environment, particularly the south Texas beaches and marine life. John

Robinson believes it will take "several years" to assess the impact fully. "We have already observed impacts," he has said. "The shore bird population is down considerably from what it's been in previous years, but I think we're only seeing the tip of the iceberg."

There are no easy lessons to be learned from Ixtoc-I. Mexico is understandably eager to develop and sell its oil, which amounts to roughly one-quarter of its gross national product. The United States, just as eager to reduce its abject dependence on Middle East crude, wants to buy that convenient Mexican oil as well as develop its own offshore resources. Despite these needs and the comforting assurances of the oil companies, however, it is becoming increasingly clear that tanker spills and offshore blowouts pose grave risks.

As Eric Schneider, a research fellow at the Center for Ocean Management Studies at the University of Rhode Island, wrote in February 1980:

> At a symposium in January at the renowned Marine Biological Laboratory in Woods Hole, Mass., the leading American marine scientists presented recent data and analysis that firmly countered studies supported by oil companies who sought to minimize the impact of oil on marine ecosystems. An assessment of the Amoco Cadiz spill shows that there is no existing technology that can fully, satisfactorily control a large-scale oil spill and that when a tanker's oil came ashore it wreaked havoc on onshore and offshore ecosystems.

The dilemma is that nothing—except doing nothing—can eliminate all the risks from offshore drilling. The real issue for the energy-starved American economy in the 1980s or 1990s is how much risk is tolerable. That is the question that should be examined closely by the American people before we allow the oil companies to proceed with accelerated offshore leasing and drilling. It is too important an issue to be left to the federal courts or the discretion of federal and state civil servants and representatives of the petroleum industry.

Chapter Ten

The End of the Road

The energy outlook for the 1980s and 1990s is not encouraging. Despite an oil glut, OPEC met in Indonesia in December 1980 and raised its prices yet again. Although production exceeded consumption, there was a market for every barrel OPEC consented to produce because consuming nations were hoarding stocks against the threat of another crisis. Any modest interruption of supply, deliberate or not, would be enough to send buyers scrambling. There is every reason to believe that prices will continue to rise and Americans will pay more and more for petroleum products. According to a 1980 study by the Congressional Budget Office, the lowest likely price for a barrel of oil by 1990 will be $84.

In this forbidding economic atmosphere, the oil companies nonchalantly continue to collect ever higher profits. One company has so much money that it literally doesn't know what to do with it. This is Standard Oil of Ohio (Sohio), whose story illustrates the difference that a bit of luck can make in the oil game.

In 1969, Sohio was short of crude and was earning modest profits as a regional refiner and marketer. Then it bought the undeveloped acreage owned by British Petroleum in Alaska's Prudhoe Bay in exchange for a special issue of common stock designed to eventually give BP 53 percent ownership of Sohio. Ten years later, Sohio's profits had jumped 2,200 percent to $1.2 billion.

This phenomenal increase came about because Prudhoe Bay turned out to be an embarrassment of riches. By 1980, 1.5 million barrels of Alaskan crude oil was being pumped through the Trans-Alaska pipeline every day, and Sohio owned 53 percent of it. Sohio has no overseas operation and very little promising exploratory acreage in the lower forty-eight states. The Alaskan oil is the company's only major asset, but since it is draining away (more than half of it will be gone by 1987), Sohio is scrambling to buy assets to replace it. As *Business Week* observed: "The very speed of Sohio's rise to Croesus-like wealth has caught its managers in a maelstrom. Trained to operate a company of decidedly shorter horizons, they have only general notions about what to do with their riches and have implemented no plans to maintain their magnitude."

At a time when most Americans were caught between a recession and inflation and plagued by rising energy prices, the managers of Sohio were faced with the problem of how to invest their mountain of cash quickly. Should it go into synthetic fuels, coal, chemicals, or into diversification outside of conventional energy, where the managers have little experience? Spending the money on projects other than developing oil and gas might stir the wrath of Congress, especially since Sohio is majority owned by BP, which is owned by the British government.

Nevertheless, that is exactly what Sohio did. In March 1981 it bought the Kennecott Corporation for almost $1.8 billion in cash. Sohio's purchase was for $62 a share (when Kennecott was selling for $27).

Kennecott is a major copper producer that needs money to modernize its aging mining facilities, which have caused high production costs at a time of slack copper prices. Sohio, which had cash and marketable securities totaling $3.82 billion as of December 31, 1980, needed to find a profitable, long-term outlet for some of that money.

Why, you may wonder, didn't Sohio use its cash to look for oil and gas in the lower forty-eight states? Sohio's managers know that the end is approaching for the oil game. Robert Levine, first vice president in the energy group of E.F. Hutton & Company, summed up the situation: "Sohio has this cash flow and it's trying to put it to use outside of oil. They want a natural resource with a longer life span. Prudhoe Bay declines in 1986. What lies beyond that? This is part of the answer."

The Seven Sisters are still prospering, although they, too, are faced with a problem—one that illustrates how times change even in the oil game. Until 1973, the Sisters dictated the price of oil after negotiations with the thirteen member states of OPEC. That was the year the Sisters told OPEC that they would not stand for a price increase from $3 a barrel to $5 a barrel. By the fall of 1980, the price of OPEC oil averaged $31.50 a barrel and was still climbing. OPEC had taken complete charge of pricing and the multinationals could no longer even be certain of securing enough oil to fill their own needs and long-term contracts to their customers, as OPEC cut back production and began to consider selling to some new customers. Some of these were as dependent on OPEC oil as the United States—for example, Japan and state-owned national oil companies like West Germany's Veba A.G., the French Société Nationale Elf-Aquitaine and Compagnie Française des Pétroles, and the Italian Ente Nazionale Idrocarburi (ENI). Before an OPEC meeting in Caracas in December 1979, Marcello Colletti, planning director for ENI, commented: "Oil is a political commodity now. It's not something to be left to markets and businessmen."

Nevertheless, the businessmen who run Big Oil had no cause for complaint about their 1980 profits. They continued to soar, although falling short of the triple-digit rises of 1979. The oil companies remained the leading profit makers among all U.S. industrial companies, with profits for their sector increasing more than 25 percent in 1980.

As usual, Exxon led the pack, reporting net revenues, or profits, of a record $5,660,000,000, a surge of 31.8 percent over 1979. Mobil racked up net revenues of $3,278,000,000, an increase of 63 percent. Texaco posted net revenues of $2,642,500,000, up 50 percent; Standard Oil of California, net revenues of $2,401,000,000, up 35 percent; Standard Oil of Indiana, net revenues of $1,915,300,000, up 27 percent; Atlantic Richfield, net revenues of $1,651,400,000, up 42 percent; and Shell Oil, net revenues of $1,541,000,000, up 37 percent.

For other players in the oil game—companies not so well known to the public, nor earning billion-dollar profits—1980 also proved to be a good year. Schlumberger Ltd. (pronounced "slumberjay" by roustabouts), with headquarters in New York City, is the world's leading supplier to oil drillers of "wireline logging," an essential service for monitoring oil wells that was first developed in France in 1927. Schlumberger earned profits of $994,300,000 in 1980, an increase of 51 percent over 1979.

Since the decontrol of oil prices in January 1981 as an incentive to exploration, the independent wildcatters are out in even greater numbers than they were under partial decontrol. In 1980, 32,900 new wells were drilled, a 19.2 increase over 1979, and the rewards for new discoveries can only increase the pace of drilling.

No one can be certain how much oil this drilling will find, but the stakes are high. However, few of the independents can expect to match the success of Mr. Wildcatter, as Marvin Davis is known in the industry. He is the fifty-five-year-old president of the Davis Oil Company of Denver, who stands six feet, four inches tall and weighs three hundred pounds. In the past quarter century, Davis Oil drilled about 10,000 wells and about 1500 of them proved to contain oil or gas. Along the way, Mr. Wildcatter amassed a fortune estimated at more than a billion dollars in oil, real estate, and banking.

Well below this financial level, the oil game looks good to certain students and workers. Graduates of schools such as Texas A & M and the University of Tulsa, which have well-known petroleum geology programs, are being offered entry-level jobs paying $20,000 to $25,000 a year. A petroleum geologist with a sound track record can easily earn $100,000 a year, with a piece of the action often thrown in.

Skilled workers are in demand and are being paid accordingly. The lowest roustabout on the offshore drilling rig of the Gulf Oil Company of Nigeria located off the mouth of the Escravos River earns $2,000 a month tax-free for less than six months of work a year, with free flights to and from wherever he calls home. Head drillers in the Rocky Mountain Overthrust Belt can earn $50,000 a year.

So while those in the oil game, from the members of Exxon's management committee to the sweat-stained roustabouts listening to Dolly Parton cassettes on a drilling rig off Nigeria, are doing just fine and can only do better under price decontrol, the consumers of oil products continue to be stretched on the rack of higher prices.

Although he pledged the total decontrol of domestic oil prices during his campaign, President Reagan did not invent the concept. The Carter administration had planned to begin total decontrol on October 1, 1981, to stimulate production and lessen the dependence on imports. President Reagan simply advanced the date to late January 1981.

The oil companies immediately began to increase the price of gasoline and heating oil by several cents a gallon. Gasoline prices could go as high as $1.60 a gallon nationally from the average price at the beginning of 1981 of about $1.26. The sharp rise in heating oil prices prompted Richard Kessel, leader of Long Island Consumer Action, to send a telegram to President Reagan urging him to declare New York a disaster area. Kessel's group maintained that heating oil prices rose by 17 cents to $1.23 a gallon in two months since December 1; the forecast is for further increases nationwide.

One can only hope that the oil companies will use their resulting higher profits, even when somewhat diluted by the windfall profits tax, to arrest the decline in domestic production. There is no certainty that they can.

What is certain is that President Reagan's approach to the oil game is to encourage the private sector to get the job done and lessen government regulation. His choice for energy secretary, James Edwards, an oral surgeon and former governor of South Carolina, has said, "I'd like to go to Washington and close the Energy Department and work myself out of a job." The new secretary of the interior, James Watt, quickly stirred the ire of environmentalists by proposing to auction oil- and gas-drilling leases for four areas off the northern California shore.

There remains, of course, one critical oil problem that President Reagan can't really solve. The Iran-Iraq conflict, while it had a minimal effect on U.S. supplies, was a reminder that supplies from the Persian Gulf, in an area close to Soviet borders and 8,000 miles from Washington, are subject to political vagaries beyond anyone's control.

Meanwhile, the greatest transfer of wealth in the history of the world continues, from the industrialized West to the members of OPEC. Saudi Arabia, which accounts for nearly one-third of OPEC's output, alone takes in $166 million a day for its oil. If its flow were drastically cut for any reason, there would be severe economic depression in the West.

In contrast to this grim prospect, the oil game did manage to provide one piece of seriocomic relief in the waning days of the Carter administration. Paul Bloom, the investigator of violations of federal oil-pricing regulations for the Energy Department, was dismissed as special counsel by the incoming Reagan administration. But one day before he left office on January 19, 1981, Bloom took it upon himself to distribute $1 million each to the Salvation Army, the National Council of Churches, the National Conference of Catholic Charities, and the Council of Jewish Federations. The charities were to use the money to help poor people pay fuel bills.

The $4 million was part of the $280 million overcharging settlement reached with Standard Oil of Indiana (Amoco). Upon learning of Bloom's philanthropy, the new energy secretary, James Edwards, tried to get the money back, contending that Bloom lacked the authority to make the gifts. Eventually, the four charities agreed to return a quarter of the $4 million, and the Energy Department agreed to drop its efforts to have all of the money returned. But many applauded Paul Bloom's grand exit from the Energy Department as Robin Hood. Commented Robert Pope, pastor of the Pascack Reformed Church in Park Ridge, New Jersey, "Mr. Bloom's imagination hasn't failed him or his heart. Yet even more, his wild act of charity reminds us of that line from *Man of La Mancha,* which Mr. Bloom seems to have taken to heart: 'Too much sanity may be madness. And maddest of all, to see life as it is and not as it should be.' "

The lingering suspicion on the part of consumers that the oil companies had been ripping them off when prices were controlled is unlikely to get a definitive answer during the Reagan administration. The new administration's budget, sent to Congress in March 1981, proposed to cut the annual allocation of Bloom's former division from $35 million to $6 million. The division had accused thirty-three of the thirty-five largest refiners and oil producers of price overcharges and other offenses totaling $11 billion since it was formed in 1977. Fifteen of the companies agreed to make cash payments of $550 million to consumers and the federal government, after which the Carter administration dropped $3.5 billion in claims against them. However, Exxon, Mobil, and Texaco refused to settle at all.

"These cutbacks would constitute a death blow to any credible effort to pursue prosecutions against major refiners," Bloom commented. Democratic Congressman John Dingell was more outraged, saying: "This adds up to amnesty for the oil companies and the public be damned." But the new energy secretary, James Edwards, denied that there would be an amnesty and said that the prosecutions would continue even under the reduced budget.

Despite the increased rate of drilling, there were no significant petroleum discoveries in the United States through the first half of 1981, but then there were no shortages, either. Gasoline stocks in the United States were nearing record levels and this, coupled with reduced demand, made it difficult for gasoline retailers to pass on the wholesale price increases of nearly 10 cents a gallon declared by the major oil companies since the price decontrol of crude oil.

But that was only the short-term situation. The adequacy of the West's energy supplies will continue to be exceedingly uncertain as long as any one of many oil-producing nations can create shortages by reducing production. The United States and Western Europe remain vulnerable to a selective oil embargo and diplomatic blackmail. Looking to the future, Walter J. Levy, probably the world's best known and most respected oil consultant, offered this gloomy scenario in the summer 1980 issue of *World Affairs*:

> We will probably be confronted by a series of major oil crises which might take any or all of several forms: fighting for control over oil resources among importing countries or between the superpowers; an economic-financial crisis in importing countries; regional conflicts affecting the oil-producing area; or internal revolutions or other upheavals in the Middle East. At best, it would appear that a series of future emergencies centering around oil will set back world progress for many, many years. And the world, as we know it now, will probably not be able to maintain its cohesion, nor be able to provide for the continued economic progress of its people against the onslaught of future oil shocks—with all that this might imply for the political stability of the West, its free institutions, and its internal and external security.

Whether any of these crises actually erupts, one thing is certain. The oil game is not yet over, although slowly but surely it is moving toward the end of the road. The United States must get on with serious conservation and the development of other energy sources.